WORLD OF HI-TECH SALES: RULES AND REALITIES

John,
Heather sends her best!
Hope you enjoy the book.
Regards,
Bob Davis
9/9/97

WORLD OF HI-TECH SALES: RULES AND REALITIES

by

Robert Davis

ISBN 0-9656094-0-5

Cover graphics designed by Phyllis R. Ross.

Inquiries may be sent to:
World of Hi-Tech Sales
1153 Route 3 North, Suite 84
Gambrills, MD 21054

Ordering information is in Appendix A.

Dedication

I have learned and am still learning a lot from my two sons, Bob and Blaine. Their being and love has taught me the importance of patience and an appreciation for life. It has taught me to look deep inside for understanding myself. Bob gives his love to all unconditionally. This is an endowment few of us will ever attain. His capacity for love is his virtue, admired by all who have met him. Blaine has wit and charm which he has developed to an art. He often asks questions which show his unique perspectives.

I thank Regina Davis for her love and support. Her confidence in me and encouragement make sure I do not slow down. You may decide to share this book with your spouse so he or she may finally get to see beyond the elegant lunches and exciting business trips. It is time they heard the story.

I thank Debbie Davis for her love and dedication. She has continually put aside her needs to ensure the well being of her progeny. Her devotion allows me to make a greater contribution.

This book is to honor all hi-tech peddlers who manage to bring in the revenue year-end-and-year-out despite their lack of resources and confused management. Maybe one day all employees will understand your importance to the viability of the company.

This book is for senior management who are not sure how to get the salespeople focused. Every stockholder and employee deserves an effective and professional sales organization. It is your job to ensure this happens! May these insights get you beyond the smiling faces and "yes, sirs" or "yes, ma'ams."

This book is for all of those people who work in jobs supporting sales people. Maybe it will help you understand how important your job and their job is to the overall success of the company. Please give a copy of this book to your manager.

This book is for students of business who sense the importance of marketing and sales to a company's well being. It may lead you to a fascinating career. There is no better way to learn about business than through sales!

This book is for all those employees who have good ideas to improve business conditions for their company but management will not listen. Hold on to your thoughts and ideas. You may one day put them into a book.

Last, I thank my parents, Robert and Anna, for their patience and non-failing support over the years. As Robert has often said, "A family's love is all we truly ever have."

Contents

Introduction

Many unemployed white- and blue-collar workers today do not understand why their layoffs occurred. Yes, they heard the speeches and saw the memorandums from management, but still it did not make sense. What happened? Where should management have been focused? The business press is swamped with articles about global competition, foreign exchange rates, underemployment, technological change, mergers and acquisitions. However, while these subjects are pertinent, a primary force driving the "Angel of Layoffs" is closer to home – the sales department.

Top management must stop pointing fingers outside of the company, towards the economy or the competition. Examine sales management in your firm carefully and you may find the source of a major corporate ill. Sales managers who do not command respect and loyalty cannot lead in tough times. They will not be able to foster an environment of teamwork and extra effort at the end of the quarter. Top management must immediately begin to analyze this function's leadership to ensure the organization is not drifting down a spiral path from which recovery gets harder and harder. Most companies' business problems can be traced to their marketing or sales organizations. These two functions are the most important corporate functions in terms of a company's sustained growth and long-term viability. This book will focus on the sales function. I will generally leave the broader subject of marketing to another day.

These are the experiences of a high-tech salesperson, me, who has been in the computer industry since 1968. I have made mistakes during my career; I did not always see the warning signs ahead. But eventually a seasoned professional emerged. I have worked for several information technology companies and seen many styles of sales management. I have been a sales manager myself, working with some of the best and some of the "amazing." I have hired and fired people, and been hired and fired myself.

A variety of incidents, presented as vignettes, are described in this book. The vignettes, presented in an encased box, are real events. However, this book has no intent to embarrass or offend anyone. Names have been changed to ensure the anonymity of all characters. True customer situations can be bizarre, some even too bizarre for this venue. All in a day's work for a "peddler."

This story is shared for all to learn the critical importance of the selling function in all organizations. This book is written as a compendium of rules, essays, and commentaries. This is not a book about Sales 101; it is a book about the insights of a high-tech sales professional.

Preface

Up and at 'em

The alarm went off at 5:00 A.M. Was it part of my dream? No... It was real. Hadn't I just gone to bed? Just five more minutes, please... but I had an important breakfast meeting across town. My customer, Mike, a former Marine officer, was always punctual and after waiting weeks to see this guy, I knew I had better not be late for our 7:30 A.M. appointment. I moaned, I groaned – but I got up.

During the drive, I began to consider which questions I would ask Mike. I needed information to find out if my company's computer systems would be included in an upcoming electronic data interchange (EDI) pilot-system evaluation being done by Mike's company for his government customer. The evaluation criteria were not clear to me. I was not sure I knew anything about the competition. I was unclear about the required timing and schedule for award and computer system deployment. I did not know who would be the key users and beneficiaries of the new system. The more I thought about my questions, the more I realized I didn't know....

How the business world has changed.... We were supposed to be working only four days a week by the 1990s, as I remember hearing in the late '60s. What happened? The work environment today seems so much harder. Back in the '70s there were lots of orders. Customers could not get enough hardware and software. Big territories, hot products...nice commission checks! But now, just to earn the same level of income I always made seems harder and harder. My manager, George, has been nagging me lately about revenue – profitable revenue. I told him, "I hear you, I am your best closer; let the bean counters worry about profit." Wrong thing to say! Everybody is so serious these days.

What time is it? 7:10 A.M. Good...I like being early. Anyone who thinks sales is an 8-to-5 job clearly does not understand the sales job. To be successful, you work until the job is done. Most situations in sales cannot wait. Some weeks seventy hours, some weeks forty hours...it is not just the quantity; it is also the quality. Focus on the customer – not the clock.

Now that I am here, I will get a table where we can talk. You never know who is listening. I remember a colleague of mine won a large order last year by overhearing the competition talking in an airport lounge about their technical strategy. They still do not know why they lost that deal!

Oh, I meant to bring our EDI training brochure with me to show Mike our EDI consulting services capabilities. Oh, well, look at the positive, it will give me an excuse to see him again. I do not like to provide too much information on a single sales call. I try to make sure there is always a reason to get back to the customer. Keep the door of communication open!

Time check... It's 7:35 A.M. I am sure Mike is just caught in morning traffic....

I have got to understand how the product evaluation is going to be done by Mike's company. At this point, I have no idea what level and types of resources are necessary to make sure my company responds appropriately. But that's my job — to find out the needs, develop the strategy, and ensure we pursue this opportunity positioned to win!

Nice thought... but the way things are at work, asking for resources is almost comical. The internal battles required to do the job right are harder than the battle to bring in the business. It all seems absurd to me. I know what we have to do to win. Why does George get defensive every time I ask for resources? He must not have confidence in me.

Time check... 7:45 A.M. I hope Mike shows up. He has missed two meetings, which were firm appointments, during the past year. When I was new in sales, I was taught to call ahead and reconfirm the meeting prior to the appointed time. I never do this. I always go on the call. If the customer does not arrive, I use it to my advantage. Customers usually feel so bad for missing the meeting that they go out of their way to help me at the next meeting. The customer is then so much more willing to answer my questions. I have gained significant insight and account knowledge from this technique. Previously when Mike missed a meeting, he apologized sincerely, but it makes me wonder.

What time is it? 8:00 A.M. It looks like he will not be here again. I will wait a bit longer to make sure I do not miss him. It is difficult to schedule appointments with him. Somedays I wonder about this job....

Davis' Rule One

Never waste time waiting for excellence from mediocre management.

Hello!

My name is Robert Davis. I have been in the computer industry since 1968. That year I touched my first IBM 360/20 during a computer technology course I took at Automation Institute in Wheaton, Maryland. I found myself shortly thereafter repairing Philco-Ford 1000 computers in the U.S. Air Force. (Yes, Ford actually made second-generation computers in the late '50s.) In the 1960s when Uncle Sam came calling, you had to make some life decisions. Back then the system helped you make these life choices... After my tour of duty in the U.S. Air Force, I attended Virginia Commonwealth University (VCU) in Richmond, Virginia as a management major with a concentration in economics. I did well in my studies and enjoyed both disciplines. Economics was interesting and helped me understand, to a large extent, the logic of the business world. Management as an academic subject was clear and logical to me.

At VCU I learned two key crucial management principles that have stood the test of time and remain basic tenets to my job success:

1. Make decisions.
2. Follow up.

Fred Haas, Ph.D., former professor at VCU, explained often that management's greatest responsibility is to make decisions, move forward, and be willing to stand by the results of your decisions. If you cannot make a decision, you are not a manager. My experience has taught me this is true.

Neil Humphreys, Ph.D., who at that time was an associate professor in the Information Systems Department, made it clear that a professional always follows up, stays in touch (SIT), and builds a network of friends and business contacts. This gospel has served me well, very well.

I was also taught, once you knew how to plan, organize, direct, control, and staff, you would be a good manager. "A manager is there to help you grow professionally and, most of all, be a leader...set an example." Following graduation, was I in for an education!

* * *

Management Styles

During my twenty-one years in this industry since college, I have worked for twenty-two different managers. I worked for one manager eight of these years; therefore, I spent an average of seven months working for the rest of them. Three of these managers I would like to work for again; four of these managers laid me off; and two of the managers and I did not get along. The rest were...there. In my many encounters with management, I have uncovered a wide range of styles.

Have you ever worked for someone who was rich or semi-retired? I have worked for managers who fit into this category. It worried me to work for someone who did not need the money. I had trouble understanding why they would put up with all the chaos.

I have worked for managers whom seemed to be in retirement — who did not rock the boat, went with the flow, and so on. I submit, to win a significant new account a salesperson has to raise hell in most companies.

I wondered how bad it would have to get for these managers to voluntarily leave the company while still having a measurable pulse.

Have you ever worked for someone determined to succeed? I have worked for hungry managers. Hungry to succeed; hungry to move up in the company. Yes, they could be too hungry and achieve success no matter the impact on those around them: co-workers or family.

But, what motivates these sales managers or salespeople to hang in there year after year? Money, stock options, the annual trip to Hawaii, are among the reasons. In some cases, these are the motivating factors, but to develop and maintain a successful, vibrant, and growing sales organization, an additional motivation is required: the satisfaction obtained by winning competitive business.

To have confidence and know their strengths, senior salespersons or sales managers must have the BIG win in their careers – they must have won the deals said to be impossible by their management. They have to have slain the competitive dragon almost on their own. I had two BIG wins in my career. I won two large competitive deals management said could not be done. They were wrong. These were the most gratifying experiences of my sales career. This is why I like sales. I like competition and I like winning. Yes, I lost some deals along the way, but rarely did I lose when I went to the wall for a win. I am not intimidated by competitors. I respect them, observe them, and take them seriously. I will not give up on a deal that is my kind of business – a deal which belongs to me! I am a *peddler*.

A *peddler* can hold any position within the organization: sales management, direct sales, or even sales support. *Peddler* is a style and a focus, not a title. A *peddler* thrives on competition. *Peddlers* are there year in and year out because winning gives them tremendous satisfaction. Beating the competition after a long, dog fight is their greatest trophy. Always has been; always will be. In a job that demands competition for success – sales – peddlers are critical to the success of the organization.

Another style often seen in companies is the *administrator.* Again, an *administrator* can hold any position within the organization: sales management, direct sales, sales support, or sales support management. The *administrator* is well aware of the internal corporate politics and strives for upwardly mobile positions. The *administrator* has an internal, organizational focus and is usually uncomfortable with the confrontational requirements to bring home the BIG win. An *administrator* should not be in the front line of the sales process where a keen eye on the competition and a willingness to "fight the good fight" are the keys to success. Appropriate administration and attention to the details are also critical to the success of the organization. Individuals with an "*administrator*" style can be well placed in another part of the organization that would benefit from the extensive reports, detailed memos, and exacting style of the *administrator.* Are your administrators properly placed in the organization to add value to the sales process? Or are your administrators getting in the way?

The administrative sales manager will frequently spend time trying to find out what is happening in an account, demanding to be taken on sales calls. A senior salesperson may be viewed as "hard to train," shying away from team efforts or unwilling to share information about their accounts with an *administrator*.

When all is said and done, the sales manager, or any manager, must command respect from subordinates. The manager must be viewed as a sales resource adding value to the sales process. They are not paid to fill out forms and attend (internal) meetings. These activities are the means to sales success, not the end. If the manager is not respected and accepted as a leader, then he or she is not a part of the sales team. However, salespeople will give that extra effort after 5:00 P.M., or on a Saturday or at the end of the quarter when a relationship of trust and teamwork exists with their sales manager.

Peddlers view their accounts as their territory, almost children, so they think. They take management they like or respect willingly to see their "babies." Other management seems to find it difficult to get into their accounts, even after repeated requests. I wonder if they will ever figure out why?

By placing sales administrators in first-line sales management positions, upper management is losing competitive sales opportunities and wasting valuable salesperson-hours. There seems to be a belief that... maybe, with the right organization chart, with another mandatory training course and enough time in the job, the numbers will somehow roll in. Wrong!

* * *

Taste of Victory

There is one key variable which is often overlooked or underemphasized when upper management selects a sales manager: sales leadership. Leadership in the world of hi-tech sales means being successful, winning new business consistently. It means beating another company in a head-to-head competitive fight. This quality of success, which is often overlooked, means being proven under fire; this separates sales leaders from administrators. How many sales managers have actually won large competitive deals in their careers as salespersons – deals they made happen largely on their own? Many sales managers harangue their salespeople to work hard and win, and they themselves have seldom tasted the sweetest victory of all – beating a fierce competitor in a real dog fight.

When I worked for George, a regional sales manager, this was the main quality he sought in his subordinate managers. Do they know how to win? Do they like stomping a competitor? Popularity and personality aside, do they know what to do to close a deal? If yes, their sales skills will command the respect of their subordinates. This notion is a basic tenet of selecting or promoting sales managers.

Many factors should go into a management selection situation. Some of the factors I have seen are relevant and some are not. I have noted some disturbing trends...

- Poor managers select poor managers to work under them.
- Competent people intimidate a poor manager.
- Administrators will select administrators to be so-called managers under them. They want reports

accurate and on time above all else, for example.

Seldom do I see someone selected for his or her leadership skills. How is sales management in your organization developed and selected? The taste of victory is sweet and a true education on the requirements for winning again and again.

* * *

Relationship Building

A primary duty of sales management is to build strong relationships with customer management. Regular face-to face meetings where sales management can listen and learn are the order of the day. Sales management needs to understand their customer's business challenges to better understand how to develop and deliver competitive and winning solutions.

Consistent management-to-management dialogue is also necessary to build an environment of trust and a mutual understanding where differing business cultures exist. The larger the sale, the more critical these discussions become to ensure joint success. Customers today have many choices of technology and business partners. While standards and open systems simplify the process of integrating technology, they do not in any way ensure achievement of business goals. A company today which can continue to reduce their cost of sales without losing the executive-to-executive dialogue will continue to thrive. This is a crucial challenge!

Today many customer meetings are conducted to either

discuss an ongoing crisis or to chit-chat about general topics. Many times I have been asked to schedule a meeting to discuss a problem, ask for an order, or on a rare occasion, thank the customer for an order. Seldom has the purpose of the meeting been to listen to the customer's business problems. Seldom has the customer been truly willing to open up and share his or her real concerns and priorities. Since the customer sees a new face from your company at every other meeting, how can you expect them to open up and speak candidly? This is a serious red flag.

Building a relationship of trust is a process that takes time and consistent contact between individuals. Can your sales managers name in rank order, by revenue, their top ten customers? Can they describe the top business concern or initiative in those accounts? Can they name two senior managers in each key account? If you ask them the last time they met with key customers just to listen, would they be slow to respond? Listen closely to the answers you hear. Be sure these managers understand their role in the selling process. If not, they are not adding value to the business process.

* * *

Insecure Management

I often see management who seem to be afraid to admit they do not have all the answers and ideas. This is a sign of insecure management. Insecure management is based upon a lack of management experience and discomfort with currently assigned duties. Please note someone could have been in a management job for years and still have no noteworthy management training and development. Time-in-grade in itself means nothing here.

There are clear signs of insecure management. They often ask for input from everybody, but seldom give feedback on said input. They often form committees or task groups to make sure an issue is thoroughly analyzed. They seem to be unusually slow in making decisions based upon the recommendations of these committees. They always seek a consensus from their subordinates (and supervisors) before moving forward on an issue. They request pre-briefings on a subject before the meeting to be briefed on the same subject. They like their subordinate managers to be "soldiers" – loyal, follow orders, loyal, team players, and loyal. "Loyalty is the highest virtue" is their credo!

Insecure management must have control of their environment. They want all resources and subordinates to be a part of the imagined sales team. Success comes from working together, not rocking the boat. They want regular business reviews, meetings, reports and procedures in place to maintain control. Too much is at risk to leave the throttle open. Insecure managers have trouble letting go of the throttle. Some have a need to micromanage to increase their perceived control. They cannot seem to let go of their resources or assets. Delegating real authority is difficult for these managers. They must keep tight rein on their fief. They also keep their real feelings close to the vest. They look for another agenda behind the stated agenda. People may have let them down in the past, so they must be careful about revealing their real feelings concerning a business issue or a working relationship in the office. They must never appear vulnerable from a political standpoint.

Do you believe your sales management will be successful? When they have been provided solid products, competent technical support talent, competitive prices, agenda, professional sales collateral (literature), and advertising, are they being successful? If not, why not? Are your sales managers providing leadership or fear to their salespeople?

* * *

Professional Management

What makes a good manager? What makes a good sales manager? Why are good managers so rare? When you get past principles of management, such as planning, organizing, staffing, controlling, and directing, what is important? Let us consider these skills as being critical:

- Politically savvy – They know when to bend and when to stand; they put substance above form. They never treat policy as a god.
- Decision maker – Willing to "fish or cut bait;" they never wait for or expect perfect information.
- Ethical – They will not consider or engage in unethical conduct. They expect their people to do what is right all of the time.
- Communication skills – Good managers know what to say, and when to say it. They listen carefully and speak clearly. They pass along information which will help get the job done and filter out unnecessary

data. They know how to be a sounding board, that is, when to listen and when to act. They are personable and like interaction with people.

- Have won competitive business – A good manager has won more than one head-to-head competitive deal during his or her career primarily as a result of their leadership and tenacity.
- Respected by their subordinates – Sales management is not a popularity contest. Managers must be respected as managers by their sales team, otherwise the requested extra effort at the end of the month, for example, will not be forthcoming.
- Able to establish management level relationships with key accounts – They can help open doors into new accounts and keep them open during positive or negative account situations.

Classical management functions are planning, organizing, staffing, controlling, and directing. While relevant, this is the old school. Two new and critical management functions are leadership and anticipation. Leadership means setting an example, setting the pace, and being able to help get the job done, if necessary. Anticipation means being proactive, being able to envision problems or risks before they occur.

Managers must be willing to get their hands dirty; they must be ready to step in and help. They will not sacrifice their subordinates' credibility for their own betterment. Leadership adds value to the business process. Many sales managers have lost sight of their true role in the selling process and have become master administrators. They are internally focused and attend many meetings which often have nebulous agendas and dubious results.

They say "no" many times more than they say "yes." They are not risk takers. They wait for those around them to make decisions. They forget their duty is to lead or guide, always staying focused on the organization's mission.

Anticipation is a critical management function. Those who have the vision to anticipate business problems and requirements are the captains of industry – the real leaders. Identifying a business trend or a future event which can be translated into a current business opportunity for your company is the primary role of senior management. It is also the role of every manager.

Can you imagine being ready for the unknown? Can you imagine having work teams and resources aligned to accept ever-changing requirements? We must begin to think in terms of opportunities not markets. Ten-page performance appraisal plans, rigid one-year goals, salary-only compensation plans, and hierarchical organizations are being left far behind. This readiness for constant change is unnerving and challenging at the same time. Sales management must be agents of change – constructive change. Managers must seek and identify ways to foster positive change in their organizations. They must find creative ways:

- of finding new accounts
- of finding new business in established accounts
- to beat competitors consistently
- to lift morale and build a consistent selling machine
- to identify and communicate company discriminators and successes to their entire sales team
- to recognize successful peddlers

- to overcome the distraction of constant organizational changes
- to show non-sales management the added value of sales management and salespeople
- to provide advanced sales training even when training budgets are being slashed.

Management able to lead and anticipate will make the job exciting and rewarding. Managers who are glorified administrators will make it demoralizing and overwhelming. Let us divert here and visit reality.

Everybody's Friend

Henry was one of my favorite salespeople. He was always pleasant and said yes to any request. Everybody in the office liked Henry. I never heard any complaints from customers. However he had trouble making quota each quarter. During our frequent account strategy sessions, Henry accepted all my suggestions and concerns.

Over a six-month period I finally realized Henry was not following through and getting results. In spite of our friendship, I knew I had to become proactive and get him focused on his duties. There was a large bank in Maryland he was assigned to penetrate. I worked with him to help develop an account penetration plan to include creation of customer presentations, scheduling executive management visits, sponsoring free technology seminars, hosting tours of similar application environments, and so on. Nothing happened. The plan was never executed. No presentations were prepared. No meetings or seminars were scheduled. Henry seemed to have no interest in penetrating this prestigious account. He continually agreed with my increasing criticism of his poor performance. I felt it was now time to take action.

I wrote a two-page memorandum to my manager detailing Henry's lack of performance and my multiple attempts to communicate my concern to Henry. After my manager received this memorandum, he requested that he and I meet to discuss the situation. As soon as the meeting began, my manager praised my initiative and commended the detail in the memorandum. Clearly, Henry was not doing his job. Then my manager explained to me that Henry and our Vice President were close personal friends and had been since high school, therefore Henry was to be left alone. Did I have any questions? I said, "No." My manager thanked me for my time and the meeting ended. I went back to my office to ponder the situation.

Never waste time waiting for excellence from mediocre management. It is difficult, almost impossible, for mediocre managers to rise above politics and complacency to see the longer view—to see what has to be done to move forward, to really understand that which is important and that which is petty. This disease is so rampant in some companies, they do not know what "healthy" means anymore. I have been through downsizing, layoffs, right sizing, and reorganizations many times in my career. The companies I worked for Tesdata, Data General, Gould, Sequoia, Tandem, Digital, and Gandalf, always seemed to me to be in a state of flux. These companies all had vision when they were started, achieving growth, revenue and profits. Then at some point in time, forces, primarily external, said, "You have to change your vision; you have to conduct business differently now." Management complacency in many cases allowed the train to get off the track. In some cases, years of success led to management complacency.

The old model is not working today! Pressure from stockholders, known and unknown competitors, and even new technology itself, which we help create, causes frequent organizational changes – changes in industry and application foci, changes in measurements, and eventually – changes in headcount....

* * *

Be Accessible

Expecting employees to do more with less requires management to be accessible and part of the solution, not some remote higher echelon waiting for results. Many companies will point to their open door policy, but accessibility means more than – "feel free to set up an appointment with me when you have a question." How each manager opens up and maintains those clear communication lines is critical.

I always maintain an open door/closed door policy. If my door is open, it means you may enter my office, regardless whether I am on the telephone or even chatting with someone. It means interruptions are no problem. If, however, I am in a more serious conversation, I will close the door to my office. If an employee wants my undivided attention, or wishes to discuss an important matter, then I expect him or her to close the door. This is a signal to me and other people, interruptions are only to be made if necessary.

At the beginning of the discussion I try to understand if the other person wants me to simply listen or be ready to take action at some point after the discussion. This distinction is important because it helps me listen better and avoid reacting to events.

It is my responsibility as a manager to be available to listen to employees, all employees. Managers who never have time for lunch with subordinates, but will sit at their desks while eating a sandwich and reading the newspaper are sending out a poor message. Managers who never have lunch with subordinates are a red flag. Are they hesitant to seem approachable and easy to speak with? Do they fear an unexpected direct question? Managers must be available to talk, listen and answer questions, share philosophies, and yes, even be available to defend a recent upper-management decision.

A strong manager, a leader, recognizes that the unstable and ever-changing organizational environment of today creates an even greater need to increase the lines of communication throughout the organization. It may be lunch, breakfast, afternoon tea or a walk around the building, but managers must be perceived as approachable and available.

* * *

Layoffs and Communication

I have observed industry downsizing, right sizing, resource reallocation, consolidations, strategic redirection and, that old-fashioned but reliable term, layoffs, since 1989 when the downturn in our industry became evident. After the steel industry in the '60s, and the automobile industry in the '70s, I somehow hoped our industry would be smarter, or luckier, but economic law is economic law! Too much supply means lower prices, which leads to less revenue, then less margins, less profits, and so on. New technology fosters more productivity regardless of demand. Inventories creep upward, staying ahead of demand. The laws of economics caught up with the boom times of the '70s and '80s. Buy, buy, buy; credit, credit, credit! Next year will be better than this year!

Whatever the reason for layoffs, this process is always painful and protracted. The rumors are always worse than what actually happens. Furthermore, many times misinformation and worrying contribute to more layoffs.

The following scenario actually happened to me:

February 1 – My manager told me I might get laid off after having been employed a total of 49 days by this firm. He was trying to minimize the surprise. (While I appreciated his candor how might another employee have reacted?)

February 14 – I was told I would be laid off in a few days.

February 18 – I was told I would not be laid off.

February 21 – I was told I would be offered a senior position on the new sales team management was putting in place.

February 23 – I was told I would be offered a sales position.

February 28 – I was told I had a senior position on the new sales team and should begin to develop my new job description.

March 2 – I was told I was going to be laid off.

March 4 – I was actually laid off by my manager's manager.

March 14 – I was asked if I wanted to do some consulting work for this firm since my job skills were still needed.

The "rightsizing" I was exposed to was frenetic in nature. I have seldom seen an orderly, consistently executed lay-off process. As displayed above, rumors, misinformation, politics, favoritism, and plain luck are key components of what should be a logically conceived and executed business plan.

Is the above scenario all that unique? My manager tried to minimize the misinformation and rumors during the entire process. He cared about his people; but it still hurt. It was still a frustrating experience. Management who has no strategic vision waits for leadership from the next level above them, and in react mode will exacerbate the pain and chaos of layoffs.

Clearly, since February 1, my productivity as an employee was low, to say the least. Multiply my experience across a department or division of a company and then imagine the level of lost productivity. The downward spiral will continue at an increasing rate.

Employees in this environment are not focused on their

job responsibilities. I wish management could communicate more clearly. A person worried about his or her economic well-being or a child's health insurance coverage is not putting his or her job duties first. A salesperson worried about a paycheck needed to pay bills is not focusing on how to keep a competitor out of an account. The lack of communication from senior management during layoffs shows, in my opinion, a lack of business control and in some cases empathy for the employees. They cannot worry about employees if they are first worried about Wall Street and major stockholders.

People are afraid to speak candidly during layoffs. No one wishes to be viewed as negative or not being a team player. So in spite of management's efforts, they are often unable to get the real issues, concerns, or even new ideas on the table. Ideas which are critical to turning a company around – identifying new uses of a product, identifying new market segments, creative sales programs to win new accounts, new sophisticated sales training which is desperately needed, capitalizing on competitors' mistakes, customers' desires, and so on are not often considered in quarterly formal business reviews. These ideas are found in the trenches, from employees who spend three or more days a week with current and potential customers. Paradoxically, these are the employees who often detest meetings and avoid forums to communicate to management. Why? Because management did not listen in the past, so why would they listen now?

Management says, we are listening; we communicate to employees. They point to meetings, memorandums from headquarters <u>to</u> the field, spontaneous field visits, suggestion boxes, "open" Q and A sessions as examples to refute these concerns. I can assure senior management they are buffered by field management, who does not want their subordinates to ask them tough questions or criticize policy decisions. Some of the so-called problems faced by management in our industry have been faced and addressed by management in other companies in our industry years ago. So why is so much time spent thrashing about on these seemingly unsolvable problems? We must learn to look beyond the symptoms. It is interesting how new blood is often brought in at a senior management level to provide new leadership, decisiveness, fresh thinking, and so on. Are not new leadership, decisiveness and fresh thinking needed even more importantly in the trenches, in the field, in front of customers?

When a company starts losing market share, revenue and profits decline, and layoffs begin, the company enters a downward spiral. The time on the downward spiral can last one year or many years. As time goes by, getting out of the spiral becomes harder. It takes even more leadership and investment for management to get back to the days of growth. When employees, particularly salespeople have fear – fear of pay cuts, fear of layoffs, fear of losing benefits, they are not selling. They circle the wagons psychologically to see what management will do next. They contract a disease known as *layoff paralysis*.

The only way you can get out of the downward spiral is to sell your way out! This may require some or even all new products, advertising, procedures, managers, salespeople and advanced sales training – that is focused training, "hands-on," real world, results-oriented training. Regardless of the new programs or new blood, you must sell your way out. Layoffs and speeches from management will not get you to the goal line.

My friend Margaret, president of a company in Maryland, recently said to me, "There is plenty of business, if you are willing to work hard." This is a powerful statement overlooked in some corporate cultures. Does your environment foster and reward hard work or pay lip service to a strong work ethic? In your environment have you been able to achieve success without investment of sweat and tears? How long do you think this will continue?

Quit Yer Bitchin'!

I decided one day I was not going to take it anymore. Customers were dumping their problems on me. New products from my company were being released for shipment much later than promised. Other employees were not returning my telephone calls. I was experiencing repeated delays setting up a couple of senior management calls to visit our key accounts. The competitors seemed to have a hot new product, which was getting noticed by my customers. Enough is enough!

I went marching into George's office and proceeded to vent my spleen. George was the regional sales manager. George was big, usually quiet and appeared to act disinterested while you were talking with him. But all the time he was carefully listening and thinking. Finally, after a ten minute diatribe, I looked at George seeking a sympathetic response. He suddenly looked at me and said, "Quit yer bitchin'. You are paid by this company to sell what you have in your territory. You are paid to get the job done in spite of a lack of resources. Now get out of my office!" With that he continued reading a document which had been in his hands the entire time. I started to say something and realized I could add nothing to his statement. He had given me a clearer perspective.

Davis' Rule Two

Professional people know what they know and know what they do not know.

When Customers Ask Questions

I encourage customers to ask questions, any question. This gives real insight into their thinking, priorities, and concerns. You may never know the real reason a question is asked in a meeting, or its underlying context.

The traditional school of marketing talked about the four P's: product, price, promotion, and packaging. The current school of marketing in the '90s has this foci: customer, competition, company, and change – the four C's! A hi-tech salesperson today must study and understand their four C's. He or she must be ready on short notice to present or answer questions about competition and his or her company. All of this must be viewed understanding the real or potential impact of change – technological, social, or financial. As a sales manager, I expect salespeople on my team to be prepared to articulate for three to five minutes each of the four C's. If they cannot do this, they are not prepared to meet with executives in their accounts. They do not have a management perspective themselves. This is unacceptable today!

During the introductory portion of a product presentation to a potential customer, Mike, the key decision maker, asked, "I now do business with A and B. What do your products do for me they cannot do?" Our presenter, an experienced technical person, hesitated and slowly began to give a general answer to Mike's question. The answer was not concise, and he did not look Mike directly in the eyes when he spoke. I joined in the response to keep the tone of the presentation upbeat, but it was too late. We had not differentiated ourselves or answered Mike's question. The presentation continued for another twenty minutes, but had effectively ended.

Answer a customer's questions directly, concisely and look them in the eyes. Do not hesitate and always be ready to answer the above question. What can your product do for me that is not being done by my current vendor? Otherwise, why go see the customer?

There are No Stupid Questions

I had taken four middle managers from my key account to corporate headquarters for a series of meetings to determine if a new computer system my company was close to announcing would meet the demanding technical requirements of an upcoming U.S. Navy program which would require massive computing power. After a morning of presentations by product marketing and engineering, the customer's technical leader, Ray, pulled me aside during a break from the meeting and asked me to ask a specific technical question on his behalf. He explained that he had political reasons for not asking the question himself in front of his peers. I told Ray I would be glad to take care of it.

When the meeting reconvened, I immediately asked the person who had been presenting before the break to explain an earlier statement in more detail. Before this person could open his mouth, the national sales manager, sitting across the conference table from me, said, "That is the dumbest damn question I have ever heard!" I was speechless and before I could babble a response, Ray quickly looked at the sales manager and said, "I asked Bob to ask that question for me during the break." The sales manager instantaneously turned to the presenter and said, "Yes, what are you going to do about that requirement?" I learned that day to respect all questions, no matter how basic they may seem.

Meeting with the Customer

During the course of a typical work week, too many salespeople spend a low percentage of time in face-to face meetings with customers. "Face time" with a customer is a solid predictor of success in the business of selling.

There are many ways to "meet" with the customer: formal meetings, chalk talks, casual coffee calls, and presentations. The key to a successful and productive encounter is focus and organization. Understand why you are meeting with the customer and identify upfront what results are expected and desired.

Recently, I was asked by a key customer to discuss our opinion of a sensitive business issue with the customer's senior management without the presence of a business partner, another company, with whom we had been working in this same account. The customer planned to have five people at the meeting, all with potentially differing agendas. The meeting had the risk takers of offending some attendees and drifting into politically dangerous waters. Additionally, we were trying to get a decision from this customer about whether or not they planned to acquire the new technology we had proposed through our business partner three months earlier and did not want to do anything which could be construed as finger pointing. Last, we had forty minutes to present our viewpoint and recommendations. How do you prepare for this critical and sensitive meeting?

First, I scheduled a meeting of our sales management and technical support people to discuss our focus for this important meeting. I went to the whiteboard at the start of this meeting and asked everyone to agree on a one-sentence statement for the purpose of the meeting with this key account. Why were we going to see this customer? It took us twenty minutes to agree on the primary purpose of the meeting. I then asked for three or four specific objectives we needed to accomplish at the meeting. Objectives are commitments, agreements, or action items you want with or from the customer. It took about twenty minutes to get four specific objectives acceptable to everyone.

Finally, I asked our group what was the premise of the meeting, i.e., what would the customer gain by spending forty minutes with us? What was in this meeting that would benefit them? It took us about ten minutes to agree on the wording of a one-sentence "premise" for the meeting. Once we had put the purpose-objectives-premise (POP) into writing, it was easy to prepare the actual agenda and presentation overheads for the meeting. Now all of us were reading from the same sheet of music.

We drafted the agenda and gave the customer our POP in writing at the beginning of the meeting. This made the purpose and our objectives for the meeting quite clear. The customer also realized what they would gain from investing their time with us. Exhibits A and B give an example of an agenda and a POP, respectively. I list our attendees so the customer has our correct names and titles at the beginning of the meeting. I rarely list customer attendees because the actual meeting participants may change before the meeting begins. Furthermore, in the past I have misspelled a customer's name in spite of my best effort to avoid doing this. Misspelling a customer's name looks amateurish. Therefore, for these reasons, I do not include customer names on the agenda (unless we are talking about a formal symposium, for example).

GAN CORPORATION
Meeting Agenda
January 14, 19xx

Introductions	Ken Heynor
SGS Project:	Jean Leslie
- Scope	
- Accomplishments	
- Planned Activities	
GAN Corporation:	John Woll
- Business Partner	
- Technology Direction	
SGS Project Issues:	John Woll
- Executive Awareness	
- TIP Impact	
SGS Business goals	Jean Leslie
Follow-up:	Ken Heynor
- Technology Demonstration	
- Executive Briefing in March	

Attendees: John Woll, Director, Sales
Jean Leslie, SGS Program Manager
Ken Heynor, Consultant, GAN Corporation

Exhibit A

Meeting Agenda
January 14, 19XX

Purpose:

Review and discuss utilization of the SGS contract to its fullest potential.

Objectives:

– Assure partnership with prime contractor – ACF Corporation

– Present our role as a capable subcontractor

– Discuss impact of TIP process

– Foster executive awareness and support of SGS Project

Premise:

This meeting will give ACF management better insight into the potential of the SGS contract so they may better position ACF to address upcoming competitive electronic commerce requirements.

Exhibit B

Other activities or demands on a salesperson's time reduce the amount of time spent with the customer. Ensure these other activities or demands are kept at a minimum. However, it is not always other tasks that reduce customer interaction. Are your salespeople and sales managers people who like meeting and being with people? Is this part of their personality? I have met salespeople who are loners at heart. One person I worked with would actually go to lunch with the office gang and then get a table for one so he could be alone. He was a nice person, but did not like being with groups of people. I have met salespeople who do their jobs quietly and to the best of their ability, but they do not really like meeting new people each week and avoid being personable. Why are these people in the sales game? Someone in management placed them in the game and allows them to stay on the playing field. Do you think they will seem sincere and affable to customers? Would you give a large order to a person you "cannot read," so to speak?

How much quality face time are your salespeople really spending with customers? If you are not regularly face-to-face with a customer, you are not selling.

* * *

The Importance of Presentations

A key predictor of sales success is the number of customer presentations a salesperson gives each month. When I say presentations, I do not mean coffee calls or chalk talks. I mean presenting your messages in a twenty to forty minute presentation using 35mm slides, overheads, prepared easel pages, laptop computer graphics, or a combination thereof. A real presentation takes time to formulate and forces the salesperson to organize his or her message(s). A salesperson giving two or less presentations a month is a red flag. He or she is not sowing seeds for the future. The salesperson may also not understand the company's strengths and discriminators, and is therefore unable to articulate these points.

Do your salespeople truly understand the company's core competencies, to use the term from the *Harvard Business Review*? Do they understand the company's discriminators, that is, unique strengths and expertise? Do they understand perceived products and services they have which are not offered by the competition? Can they articulate the answers to these questions in a concise manner to potential customers?

The answers to these questions are critical in order to measure the competitiveness of any sales force. When I join a company, I immediately spend time seeking answers to the above questions. Customers will ask these questions. Why not have the answers already integrated in your presentation material? Today salespeople cannot afford to pursue business which does not match their company's strengths. Today, resources are limited and lost time cannot, of course, ever be replaced. The marketplace today will not forgive inefficiency or dabbling. Only companies with a clear focus on their target markets are going to hit the mark consistently.

It is incumbent upon management to have company and product information available to all sales (and management) people for use in presentations. It is incumbent upon the company to have a variety of media available to facilitate the salesperson's job of organizing a presentation. It is the salesperson's job to organize the information into the correct form so the presentation is concise and focused for a specific opportunity.

Some companies make preparing an up-to-date customer presentation so difficult, the salespeople have given up. Salespeople in some companies have even gotten by through convincing the sales support staff it is their job to "do presentations." Wrong! Salespersons who do not give regular presentations are cheating themselves and the company. A salesperson unable to give professional presentations due to lack of ability or interest should immediately consider a career change. A sales manager who lets his or her people go weeks without giving presentations to customers is not managing the selling process.

* * *

Heed the Signs!

When you are meeting with a customer what are the signs or indicators the customer is interested in your products or services? What are the clues that the customer is accepting what you are saying? Besides appearing to be listening, I have found these actions to be clear signs of interest when a customer:

- Takes notes carefully to ensure their accuracy.

- Asks insightful questions showing he or she is analyzing the nuances of your offering.
- Discusses his or her business problems in detail.
- Shares, to a certain extent, a personal concern or problem (but does not cross the business versus personal relationship line.)
- Extends the length of the meeting, knowing he or she has another commitment.
- Gives you plenty of time to answer complex questions.
- Unexpectedly asks an associate to join the meeting to hear your message.
- Refuses to take telephone calls and ignores interruptions.
- Insists you meet his or her manager and schedules the time of the appointment.
- Gives you evidence they specifically prepared for the meeting.

After my first or second meeting with a customer, I look for two or more of these signs to help me qualify the customer's real interest. This, of course, also helps determine how willing I am to place this new lead on my forecast.

If it Sounds Too Good to be True...

I had been working hard to penetrate a large account in Silver Spring, Maryland. This customer clearly bought my company's type of computer systems and would be considered a prestigious new customer. Various tactics to penetrate the account had been unsuccessful. I finally received a call one day from a technical manager requesting product information. I immediately went to the account armed with technical and product literature.

The technical manager introduced me to a couple of his subordinates and the meeting began. The meeting was friendly and gave me a lot of insight into a computing challenge of theirs and its concomitant concerns. They answered most of my questions. I showered them with literature and assured them this application environment was our kind of business. I left the meeting with some system diagrams we had sketched on the white board. The next couple of days were spent generating a product quotation with detailed pricing, which was over $600,000. I also developed a small proposal citing our strengths and benefits, then I called the manager where upon another meeting was quickly scheduled.

My sales support analyst and I went to the account and presented why we were well suited to address this computer processing environment. All of their questions were answered. The meeting went great. I felt good about this opportunity and knew it was ready to be forecasted. I finally had a foothold in this account!

> However, as the days went by I could not reach anyone on the telephone. No one would return my calls. The meetings had gone well, but something now seemed wrong. I called higher management, but to no avail. After further telephone calls and letters over a period of three months, I gave up. Then one day while at lunch in a restaurant in Silver Spring I met a man who worked for this company. I began to ask how so and so was doing and was their project moving forward. The person answering knew I worked in computer sales but thought I worked for another company. He proceeded to tell me they had pumped my company (and me) for data so they could attack the bid of another company who had proposed using our computer equipment. In other words, my data had given them competitive information to allow them to technically discredit my company's products. I was stunned. I did not realize there were people who would spend their company's time to collect data for this purpose.

During 1980 I was beginning to experience continuing success in my career as a hi-tech salesperson by winning new accounts and competitive deals. I got busier and busier at work.

At the same time my son began to exhibit signs of a serious neurological disorder. His mother and I were new parents and did not fully understand the gravity of his problem. Furthermore, as we began to interact with the pediatricians, neurologists, and therapists we were lead to believe his problem was "manageable" and certainly not new to medical science. During 1981, the real gravity of our son's illness slowly came into focus. We were dealing with a rare, life-threatening disorder.

However, success in the sales arena takes time, effort, and attention. Matters at home were shuffled into the back of my mind. I was planting seeds I would later reap. Also, I am not sure I was properly focused on my job responsibilities.

Looking back at my failed attempt to penetrate the Silver Spring account, I realized there were signs I chose to ignore. Maybe I was overconfident. Maybe I had other things on my mind. I had met and talked with the same people. My attempts to meet their management were brushed aside. All of their requests were verbal; they never showed me any of their requirements or concerns in writing. During our meetings, they were extremely polite and agreeable, too polite and agreeable. Offers for additional technical meetings were declined. A suggested meeting to meet my manager was not of interest. Lesson learned here. Pay attention! I invested my time and limited resources in the future more carefully.

* * *

Inside Salesperson

I needed information quickly. How was my key customer going to decide which computer system to use for a new project they were going to bid? Who would make the actual computer selection decision? How important was price, performance or software compatibility with existing deployed systems? Who would approve the recommended computer system? What was their decision-making process (DMP)?

Sales 101 teaches salespeople to develop an "inside salesperson." This is a customer who, for any number of reasons, takes the salesperson under his or her wing to help guide him or her through the DMP of the account. They may sincerely like the salesperson, dislike the incumbent supplier, want to ensure a more competitive computer acquisition, want someone to listen to their ideas, be a generally helpful person, or simply have an arcane agenda. Whatever their motivation, they are a valuable resource. In fact, I never won a truly competitive deal without having developed an inside-salesperson relationship with an employee, usually a middle manager. This manager is high enough to go to upper management meetings and get the internal management memorandums, but hopefully not too involved in upper-management, killer politics.

I laid out a sales call plan which included several questions for which I needed answers so I could develop my own detailed account strategy. In a large account, I develop two or three inside-salespeople relationships. It is important to have multiple ways to correlate information picked up on the grapevine and to not get buried into one person's agenda. For this visit, I decided to go see Don.

Getting the Information

Don was a senior project manager who had been with the company a long time. He had the ear of senior management because he was known for being candid - too candid sometimes. Don often found himself working on the most difficult and challenging projects. I viewed this as a measure of management's confidence in his abilities.

After another meeting at the account earlier that morning, I stopped by Don's office to say hi. I wanted the visit to appear impromptu and unplanned – just a coffee call. I went to Don's office and looked inside to see if he was there. Don was sitting behind a desk buried in books, reports, and computer printouts. An ashtray was overflowing with cigarette butts and gum wrappers. The chalk board behind his chair was completely covered with cryptic markings, the envy of any mad scientist. Two half-filled coffee cups on his desk looked untouched for days. The coffee appeared to have the consistency of chocolate mousse. The trash can beside his desk was buried under a pile of trash.

Don was sitting low in his chair looking at something on his desk. He looked tired; his eyes were red. He seemed to have a five o'clock shadow and it was only 10:00 A.M. Don looked up from his languor and muttered, "Hi." I took the opening to quickly enter his office, grabbed a chair and said, "Good morning." I asked him how things were going.

Don hesitated then slowly began to tell me about

his wife's serious respiratory disease. He was worried about her. She was not doing well. His son was also having unusual health problems. He did not know what to do. Maybe they should see another doctor. He had heard about a new drug which might be available soon. The medical bills not covered by health insurance were overwhelming. Here was a man under a lot of stress who wanted to talk - needed to talk. I sat and listened. I shared some of the experiences I had with my son. I really did understand some of Don's pain and frustration. After about forty minutes, Don became quiet. I thanked him for his time and said I just dropped by to say hello. He nodded his head quietly. I went to him, shook his hand, and said, "Don't give up." He nodded his head again. I then left his office.

I immediately went down the hall to Bill's office. You could always depend on Bill to tell you what was going on in the account – provided you agreed to play him a game of racquetball, which was his favorite sport. We played racquetball that week, I won and got the information I needed.

* * *

The Ego

An ego can be an asset or a liability in the world of hi-tech sales. It must be watched and managed, or it will pop up in places it does not belong. A healthy ego can motivate a salesperson to work harder and compete longer. It helps the salesperson keep his or her eyes focused on winning. Always take business seriously, never personally. Egos must take a back seat to getting the job done. Teamwork cannot be accomplished when personalities become paramount in an organization. Life is hard enough when you get along.

Too much ego can cause overconfidence. You may forget to do your homework, take a competitor lightly, or knock off at 5:00 P.M., believing all is well that day. Egos are here to serve, not to lead.

After I voluntarily left Data General, I went to work for Systems Engineering Laboratories, Inc. (SEL). SEL's computer had a unique architecture with powerful input/output (I/O) capabilities. Therefore, I sought potential customers who needed lots of sustained I/O in their application environment. One such company in Baltimore was developing what was then a new product, a CAT scanner. Not only does a CAT scanner need high I/O bandwidth to process x-Ray images, the company was planning to use a Data General minicomputer to facilitate this task. This obvious mistake on their part got me agitated.

A Presentation to Remember

I immediately called the Vice President of R&D and demanded a meeting to show them the error of their ways. The VP reluctantly agreed to assemble his technical team to consider my arguments. I grabbed my brochures, and support analyst, Tim, and off we went. It was time for them to hear the Truth!

Tim and I were ushered into a small meeting room with a chalkboard. After the usual introductions, I immediately launched into a tirade presenting how my system was clearly better than the simplistic Data General platform they were planning to use. I drew pictures of buss architectures; analyzed I/O and memory bandwidths; compared interrupt acknowledgment times, and contrasted each system's expandability. There were few, if any, questions from the customer. They did not need to speak; they needed to listen to Truth! Tim, my support analyst, was also speechless during my presentation; matter of fact, he seemed to be in awe.

It was about this time, after twenty minutes of a non-stop monologue, my brain began to send me signals that I was experiencing a draft in my crotch which should not be there. I took a pause while facing the customers and slowly looked down at my crotch. Not only was my zipper down, but the flesh of my manhood was clearly visible. I calmly turned and zipped up my pants. I continued on with my technical presentation without missing a beat. Tim was still agape in the back of the room and the customer sitting nearest to me kept staring at my crotch, but I continued. I concluded the presentation about fifteen minutes later. Tim and I thanked them for their time and left.

Always make sure your presentations make a lasting impression. I will tell you while they never bought anything from me, they always remembered me!

Moral of the story: Arrogance always creates a bill which must be paid. This character trait is dangerous and can bring career progression to a halt. Customers may tolerate ignorance, meaning a lack of knowledge, from a salesperson but they will never tolerate arrogance. Some salespeople and sales managers who have made a lot of money and have a wall full of sales award plaques need to take a hard look at themselves, before this is done for them.

* * *

Continuous Learning

The world of hi-tech sales is known for its continuous march of new technology. New products hit the street every month, and a new generation of technology emerges every eighteen months. The world of technology is in perpetual motion. People who have had to leave the industry for one year or more find getting back up to speed, knowledge wise, almost impossible. The day you stop or take a pause in learning, in the world of hi-tech sales, you have retired. You are telling your competitors to come take what used to be yours. As a result, learning must be a continuous process.

No company can possibly train all its people to keep up with the progression of information technologies and application environments. It is a joint responsibility of the salesperson and the corporation to ensure the salesperson has the knowledge and information required to be successful.

The onus to stay informed must fall primarily on the sales professional. There are a variety of ways to continue learning and increase knowledge. All of the successful people I know have developed their own individualized "self-development mix" to ensure they continually update their knowledge base. This mix can consist of a wide range of items such as:

- Regular subscription and review of selected industry publications (avoid information overload);
- Attendance and participation at key trade shows each year;
- Attendance at a couple of targeted seminars each year;
- Meeting with industry gurus to chat about the direction of technology;
- Careful listening to customers who are describing the business problems they plan to address using information technology – from any company, not just your own;
- Occasional trips to a well-stocked bookstore to review and buy books which present technology overviews or industry trends; and
- Participation in specific company training when available.

In my particular "self-development mix," I have developed and delivered a couple of seminars which I give to various groups about twice a year. These presentations range from one to four hours in length. Preparing, presenting, and updating this industry material keeps me sharper. It also frequently introduces me to new business contacts. This is another aspect of professional development – continually expanding your network.

The corporation also has a responsibility to contribute to the learning process. The standard training provided by corporations should consist of two major components: skills training and product training. This training must be focused, crisp, relevant, and timely.

Training people faced with the task of training hard-boiled peddlers have a difficult job. It is important to differentiate "training" vis-à-vis "learning." Training can be provided, but it does not ensure learning. Learning is a democracy. Tools that provide the *peddler* a competitive edge are eagerly acknowledged and snatched up. Are you getting the return on your training investment?

Learning can happen anywhere and under many diverse situations. The best ideas are not the final steps of progress, but always just the stepping stones to further progress in understanding. The joy of learning is not the acceptance of what others have learned and said, but the use of what others have rightly learned and said as a foundation for more learning. There is never an end position in training, never a point where learning stops and mere acceptance takes over. The professional is in a constant learning mode. The process of self education should never stop. If it does, you are marking time and your knowledge base is rapidly getting stale. You have retired!

* * *

Success Is Fleeting

What is considered a definition of success in the world of hi-tech sales? *The Random House College Dictionary* cites the following definitions of success:

1. the favorable termination of endeavors,
2. the attainment of wealth, position, or honors, and
3. a successful performance or achievement.

The above definitions imply a final result or ending after a period of time. Success in the world of hi-tech sales is elusive. At one time during my career I could display over twenty-four various sales – and sales management-related awards on the wall of my office, for example, Branch Manager of the Year, exceeding quotas, and my personal favorite – Most Tenacious Salesperson. The old cliché, "What have you done for me today?" is an understatement in our business. What I did last quarter or last year is irrelevant. When I finally began to understand this perspective I stopped displaying any sales awards in my office. I let my numbers speak for themselves. I have in fact thrown away some of my plaques and certificates. They simply do not matter now.

The Bribe

While contacting insurance companies in my territory one day, a data processing manager told me his firm was about to buy a new computer system. The manager explained that the evaluation process was being handled by their accountant. I obtained the accountant's name and telephone number then called him to schedule an appointment. An introductory meeting was scheduled at his office the following week. I was looking forward to this call because I assumed if the accountant was driving the computer selection process, instead of the data processing manager, the firm could not be too sophisticated in their usage of computer technology. Clearly I could add value to the customer in their computer system selection process.

I went to my first meeting with the accountant loaded with ideas and data. After a brief introduction, I began to describe our system's capabilities to the accountant who seemed somewhat disinterested. I then realized I had been doing most of the talking and asked the accountant to describe their requirements, anticipated applications and operating environment. The accountant looked at me and calmly said, "I do not care whose computer we buy. They are all the same to me. The firm willing to make it worth my while to recommend them to the insurance company will be my choice."

I stared at the accountant briefly and said, "Thank you for your time today. My company is interested in being considered for this new project. I am going to leave my card, brochure, and a suggested computer configuration for you to consider. If you have further interest, please give me a call." I then rose from my chair and left the room.

The accountant never called me back. I wanted to tell someone at the firm about this meeting, but would I be believed? I understand they eventually did acquire a computer system from another company. This was the only time in my career there was an explicit request for a bribe. I walked away, quickly.

Davis' Rule Three

When you are damned if you do and damned if you don't, then do it; at least you did something.

Making Decisions

My college degree is in management. Therefore, I studied various management-related subjects while a student at Virginia Commonwealth University. I studied management theory, principles, and issues. Planning, organizing, staffing, controlling, and directing were taught as time-honored management functions. We also talked about leadership and the obligation of a manager to be a leader. Most importantly, Fred Haas, Ph.D., insisted that to be a manager you had to make decisions! Right or wrong, a manager must be willing to make decisions and stand by the results of those decisions.

I agree with this and believe it is of paramount importance in sales. A sales professional who will not make a decision or waits and waits until the "time is right" erodes the tactical nature of a sales situation and helps level the competitive playing field to his or her disadvantage. It is like the difference between regular chess and speed chess. If a regular (six minutes or so per move) chess player plays a speed (thirty seconds maximum per move) chess player in a speed chess match, I know who I will pick to win. Hi-tech sales management is a speed chess match business!

Furthermore, delays in making and communicating decisions have an effect on morale. A manager who constantly delays decision making is viewed as not being a leader; no *peddler* wants to work for an *administrator*. Delays in decision making can affect productivity and ultimately success, because resources are held in rein or misdirected while management sits in neutral.

In sales, you must strike while the iron is hot! Effective and timely decision making at all levels is critical.

Salespeople often find themselves caught between a rock and a hard place. If they do what the customer suggests, they may anger someone else in the account. If they ignore a customer request, the competition is waiting in the wings to steal the deal away. Each hour that passes without action or a response can lessen the chances of winning. As information has a time value, so does taking action in a competitive sales opportunity. The longer the time delay, the less impact the action has. When you are damned if you do and damned if you don't, then do it; at least you did something. Not making a decision and taking action is the death knell of success in sales.

Moved to the End of the Line...

"Bob, where is the order? You said you would bring in that order six weeks ago. Do you know what's going on in the account?"

"Yes, boss. I'll call the customer immediately." I had no idea what was causing the delay.

I called Gene, an engineer in the account who had selected our system due to its price/performance ratio and unique I/O capabilities. I explained to Gene I needed the purchase order quickly. He said he would call Purchasing to find out what was causing the delay. Everything had been approved for this $130,000 computer system purchase by his Vice President, as far as he knew.

Two days later Gene called back and calmly explained that my purchase order was number 13 behind a stack of 12 other purchase orders for Digital Equipment Corporation's (Digital) computers lying on a senior vice president's desk in Pennsylvania. The 13 systems were being bought for various R&D laboratories throughout their corporation, according to the cover memorandum with the purchase orders. The customer was using capital funds to pay for these systems. The senior vice president was apparently irritated that the justification for buying these systems was poorly documented and was not sure all these computer systems were needed.

Clearly, it was time for a management visit to see this senior vice president to unruffle some feathers, but not from my management! After I finished speaking with Gene, I immediately called a District Sales manager I knew at Digital (a competitor!) in Maryland and explained to him an order for 12 Digital computers was sitting on a customer's desk in Pennsylvania. I gave this sales manager the name, address, and telephone number of the senior vice president in Pennsylvania and suggested a high-level manager from Digital call on this customer immediately to ensure there were no undue concerns. The sales manager assured me he would pursue the matter.

I do not know if the management call by Digital was actually made. My purchase order arrived in the mail three weeks later.

Attacking IBM...

Several times during my career I worked for management who believed the best way to win a new account was to attack IBM's installed base. My new boss told me to go find smaller IBM-installed accounts to sell our new minicomputer. So with no training and little forethought, I started calling several small IBM computer shops in my territory looking for a chink in their armor. It made perfect sense to me. My minicomputer system's operational costs were significantly less each month than an IBM computer from an operations and maintenance standpoint. So armed to my ankles with this fact, I decided I was ready to be bold.

The Road to Hell is Paved with Good Intentions!

I called the account with little preparation and asked to speak with the data processing manager. A secretary answered the telephone and again I asked to speak to the data processing manager. Suddenly, I was talking to Ms. Brown. She had a pleasant voice and agreed to see me after I told her how much money I could save her firm each month in maintenance charges alone. The telephone conversation went well and I was curious about her seeming willingness to discuss replacing her company's IBM computer system. Serious consideration for replacing an installed IBM computer system is normally a sacred matter only discussed privately with a priest.

I arrived on the agreed date and was ushered into her sizable office. The office was graciously decorated with large paintings of flowers and richly appointed furnishings. She appeared to be around thirty years old, was quite attractive and congenial. I could not believe it! In role playing in sales training, the data processing manager was never like this. We hit it off well! She liked my ideas and told me we should schedule a meeting with the president of the company, which she would arrange the following week. I told her we could plan for this meeting over lunch. She agreed. I was in sales heaven!

During the height of my euphoria, I noticed a picture of a handsome man on the credenza behind her desk. She was not wearing a wedding ring and my curiosity was killing me. At the conclusion of this fantastic sales call, as I rose to leave, I said, "I don't know who the guy in the picture is, but if he is your husband, he's a lucky man." She immediately slumped into her chair and said, "He was my husband. He was killed in a terrible car wreck last month." Then she started sobbing uncontrollably at her desk.

I slowly sat down again and did not say anything for about ten minutes. I offered her my handkerchief and expressed my condolences. After she regained her composure, I quietly left her office.

I never called her back or followed up. There must be a psychological distance between customer and salesperson. Do not cross this line. If you get too close or personal, the sales job becomes harder, even impossible.

Managing People

Managing people can be a difficult, yet rewarding experience. In the best case, a mutually beneficial relationship is established where the manager provides guidance and mentorship to an employee, and the employee provides fresh insight and information to the manager. In the worst case, both manager and employee maintain no mutual respect and exchange occasional conversation unrelated to work products. Most relationships lie somewhere between these extremes.

As a middle manager, I expect my subordinate managers to continue to go to professional management training courses and listen. I do a lot of "management by walking around" so I may see and be available to all employees. I have breakfasts, lunches, or coffee breaks on a regular basis with all employees in my group, to the maximum extent possible. I seek customer input about their confidence in my management team. I seek opportunities for dialogue away from the workplace with employees. I also expect my subordinate managers to be able to write 75% or more of their subordinates' performance appraisals. Too many times I have had to write my own performance appraisal because my boss was not aware of my day-to-day work. Then, of course, if my perspective of my job performance did not meet his expectations, the annual performance review skirmish began.

As a manager, I sit down with each of my subordinates, who report directly to me, every four to eight weeks – depending on the person's experience level, the work schedule, past discussions and have a 30-minute chat about their work, performance, and concerns. This is a two-way conversation, and I take informal notes during the meeting. I keep a file of these notes for every subordinate. At the end of the year, when performance appraisal time arrived I use these notes as a basis to write a fairly accurate performance appraisal. I then give a draft copy of the appraisal to the person for his or her review and informal comments. After this, the performance appraisal is finalized, we schedule a formal meeting to discuss his or her job performance. These discussions are low key because of the earlier ground work. I do not wait an entire year to discuss expectations and job performance with my team. This approach makes writing and discussing performance appraisals relatively easy.

Another key to appropriate management is recognizing and addressing performance issues. Signs of an ineffective salesperson include:

- disorganized – does not submit reports on time; overcommits
- poor preparation for customer-related meetings
- slow to follow up customer requests – may not be paying attention
- seems distracted – not focused
- too few customer presentations or meetings each month
- does not return telephone calls in a timely manner
- few additions or changes in the monthly sales forecast

These signs indicate that the problem salesperson will not magically improve or suddenly have a vision to get with the job program. They must be proactively managed to get on board the selling machine. Sales management has a fiduciary responsibility to every stockholder and company employee to drive this process and not overlook this matter hoping the poor performer will somehow just go away. A manager who continually overlooks non-performers sends a negative message, which in effect establishes a minimal acceptable standard of performance for the entire sales team.

Have you ever wondered how senior or mid-level executives know their subordinate managers are really doing their job as managers? This question has intrigued me for years. My favorite scenario is where a manager has supreme confidence in his subordinate manager and believes all is well. In actuality, his subordinate manager's subordinates hate their boss and only do enough work to stay out of trouble. This is disturbing. No one could (or would) convey to higher management the real nature of the situation. Meanwhile the senior vice president continues to chant about teamwork at the quarterly sales meeting.

An Accusation

I answered the telephone in my office and said my usual, "Bob Davis." A female voice said in a clear, firm tone, "Valerie is dealing in drugs. She is in trouble with the police. This is no joke." Then the caller hung up.

I sat there for a moment stunned. Who called me? Could this be a joke? Why would the caller say this? Was the caller telling the truth? What should I do?

Drugs were a subject I read about in the paper or heard about in the news. My company had no training program relating to drug abuse and I had not had any experience with this subject. I immediately went to Human resources (HR) to report the telephone call. I felt I should be on record as treating the call seriously and having reported the call to someone in HR if it became important later. I discussed the matter with Joe, our HR Manager, who listened and said little. He asked me to inform him if there were further developments. Joe did not provide me any other advice on the matter.

That afternoon I called Valerie into my office. I had decided to treat her as I would like to have been treated had I been the target of such a call and to give her the benefit of the doubt. I told her what had been said to me. She appeared to be totally surprised. I asked her if there was any truth to the allegation. She said absolutely not and became upset about the matter. After she regained her composure, I thanked Valerie for her time and sent her on her way.

The telephone call upset me, because the tone of the caller was quite convincing. I began to observe Valerie more carefully. I began to notice some disturbing trends. Valerie, who was single, and had no dependents, no mortgage, or car payment, for example, was under constant financial pressure and often asked for a raise or increase in her draw. She was frequently late to meetings and appointments, but always had an excuse. Valerie seemed to have wide mood swings while at work. She could be laughing one moment, cursing the next. Her co-workers were either close friends or people she "hated." I began to realize her forecast seldom changed. She was not really developing new account opportunities. She seemed to keep waiting for business to drop into her lap from the sky.

To address business issues, Valerie and I had a meeting for about one hour every two weeks to discuss her account plans. Was she making progress? Was she establishing a relationship with her accounts? Did she have an understanding of her account's strategic plans? Who were our competitors? We would talk during these regular review sessions and seemingly leave with a meeting of the minds and an agreed course of action. However, the next time we would meet, I would see no measurable progress. I seemed to be having continuing difficulty getting Valerie to meet her commitments.

I began to get irritated with our never-ending dialogue and during our next discussion, challenged her to present data showing specific account knowledge. She resented my intrusion into her sales activities and became defensive. Finally, when I noted that she had failed to set up a meeting with management in her key account as agreed earlier, she screamed, "You are full of shit. You do not know what you are talking about." This caught me off guard to say the least. True, I had never been to the account and did not know the customer. I decided to extend to her the benefit of the doubt (again) and give her another opportunity to get visible results in her key account.

There is a proclivity by sales managers to expect subordinates to develop a sales opportunity in the same manner they would themselves. Managers often have an idea of how they believe selling is done and if you do not follow their "model," you will not be successful. While I agree there are consistent steps and milestones in the selling process which will lead to success, expecting salespeople to run the same route to the goal line is ridiculous. Yet, how many times do sales managers judge or criticize salespeople based upon their own style of selling? No, I did not take her comment personally, but I was concerned.

Valerie's comment seemed to be the result of a lot of stress in her life. I did not like the tone of our confrontational discussion. I began to wonder if she lost control with her own manager, could this happen with a customer? Our working relationship was never the same after this meeting. We continued to meet and develop action plans with minimal results.

A couple of months later I left the company. A few months after I left, Valerie was laid off by management. I found out later she had been a heavy cocaine user while working for me.

Emotional Decision Making

People leave companies for several reasons. Typical reasons given for changing jobs include being laid off, seeking more money, better benefits, having a shorter commute, less travel, more challenge, greater (perceived) security and so on. There are sometimes deeper unspoken motivations to change jobs. I had just finished a year almost double quota ($15 million versus a goal of $8 million). I was chosen as the Sales Branch Manager of the Year. I had a solid team of sales professionals working for me. It was easy for me to envision a long rewarding career at Gould, my employer. Then something unexpected happened.

Our three-year-old son unexpectedly suffered severe brain damage resulting from prolonged oxygen deprivation. In a matter of thirty minutes, my son went from walking, talking, dressing and feeding himself to functioning as an eighteen-month-old child the rest of his life. I was numb. I was in shock. Writing about it years later is still painful. I did not know what to say or do. It has taken years to begin to mentally grasp the gravity and pain of this tragedy. I wish now I had been less focused on me and more supportive of my wife during this time. Her pain was equally deep if not greater than mine. I failed to acknowledge or grasp her anguish. I have no excuses; I was numb. I could not cope. Thinking about the future drove me crazy. I needed to be busy, extremely busy. My job at Gould did not keep me busy enough. I had to be more mentally challenged and therefore made an emotional decision (although I did not realize it at the time) to change jobs and jumped into a computer start-up company environment. The smell of stock certificates, bonuses, and power made the leap easier to rationalize. I was not running away; I was advancing my career, so I thought.

The start-up computer company I chose was based in New Jersey. They hired me to open their D.C. office and build a federal and commercial sales organization in the mid-Atlantic area. I was so enthused, I convinced a close friend, one of the leading sales support analysts in the industry, to join my staff. Soon after joining this company, I was sent to Anaheim, California, where we were debuting our new, fault-tolerant, UNIX-based computer system to the world, at the National Computer Conference that year. I literally stood in a large tent erected in a parking lot without air conditioning for five days, hawking the virtues of this new system which was going to change the world of transaction processing. We had a large booth, product literature, and technical staff who could answer any question and an incredible product demonstration with lots of flashing lights. What we did not have was a real product. I believed I was selling a real computer system. I did not discover until months later, there was an Altos Computer being used to drive our screens and the cabinets were only a wooden mock-up. I had no idea of this misrepresentation, and I was selling this wooden crate as hard as I could. Now I understand why some of my customer contacts were always willing to meet with me, but skeptical of new products from my company.

Meet you at the Bar!

"Okay, I'll meet you in the hotel bar. See you there. Bye." After I hung up the telephone, I realized I was looking forward to seeing Reggie. He was my inside salesperson in a major account and was always willing to share interesting information. He really understood the politics and decision-making process of his company's upper management. Time spent with him was always advantageous. Besides, he mentioned he was beginning a new project.

I arrived at the hotel bar at 3:40 P.M. for our 4:00 P.M. meeting to check out the place and select a corner table where we could talk privately. I had taken a newspaper with me which I began to read. After reading the newspaper and finishing the crossword puzzle, it was 4:25 P.M. Oh well, Reggie must have gotten tied up in a meeting. I began to read the menu and decided to order an iced tea. I soon went to the lobby of the hotel and brought back several travel brochures, a tourist guide, and a local map which I began to read. Time went by as I waited and mentally reviewed my customer calls for the next day. I wondered about Reggie's new project. It was now 5:45 P.M. The waitress brought me a bowl of soup and a turkey sandwich. I decided to have a beer since it was happy hour. I ate, read, and sipped my beer.

Reggie was really late now. Maybe he forgot or an emergency came up. No problem, I will stop by his office tomorrow. I decided to have another beer. I had no other plans and did not want to sit in my hotel room. After a few more beers, I realized it was 9:00 P.M. Time for me to go. I started to stand up and realized I was inebriated; I could barely walk. My tongue felt like it weighed six pounds. I got a grip on myself and focused on taking my first step towards the lobby.

Suddenly, out of nowhere, Reggie appeared and said, "Hi, Bob. Good to see you!" I pushed the word "hi" out of my mouth and dropped into my seat. Reggie said, "I am sorry I asked to meet so late, but I need you to explain the options available to provide high-speed, fail-safe connectivity between two large on-line computer systems. I am giving a presentation early in the morning and must be ready to address this capability."

Struggling to appear coherent I asked, "Reggie, weren't we supposed to meet at four this afternoon?"

"No, Bob. I said nine o'clock."

With that Reggie handed me his pen and unfolded a napkin and asked me to draw a diagram showing all connectivity options. I paused and appeared to be in deep thought. I was actually in a stupor and knew I could not draw a single line on the table, let alone the napkin. What do I do now?

I looked at Reggie and said, "Reggie, you are the smartest engineer I know. You know what the possibilities are. How would you expect the connectivity to be done?" I then pushed the napkin and pen to his side of the table. He looked at me briefly and then began to draw a complex diagram on the napkin. When he finished, he asked me what I thought. I said, "Excellent!" He asked me to begin to prepare a price quotation for the necessary hardware and software as soon as possible. It was now 10:00 P.M. Reggie said he had to go and wanted me to stop by and see him tomorrow after lunch. I said, "Excellent!" He left; I staggered to my room desperately clutching the napkin. The quote was in Reggie's hands by the close of business the next day.

Davis' Rule Four

A well executing selling machine has many diverse parts.

Teamwork

Management often speaks of teamwork – that mystical state sought for decades in corporate America. Teamwork should exist without management having to treat the word as a mantra. Real teamwork is built when there is a commonality or linkage between performance measurements across multiple functions. All team members should be focused on new account wins. Articulating that focus is a management responsibility.

Developing a team and coordinating the actions of that diverse group of people takes work, planning, and commitment. The responsibility for identifying the team members and executing the plan lies with the sales organization. The sales organization cannot work in a vacuum. Hi-tech selling is a process – a process involving the cooperation and efforts of many people. The larger the deal, the more cooperation and effort is required. A salesperson who primarily works alone without involvement by other employees is a red flag to me. Competitive deals are never won (or lost) by one salesperson with no help or involvement by anyone else. The salesperson must be in the role of conductor to ensure the customer hears pretty music, not a note here and a note there.

If a salesperson is working alone without utilizing key resources, management needs to be looking at the situation closely. Why is he or she working accounts alone? Why does he or she avoid forms of team building? Is the salesperson afraid of involvement by other people? If so, why? Is there fear of management oversight or an upper-management sudden visit? Is the individual "too senior" and should be mentoring a person new to sales? What happens when this "Lone Ranger" leaves the company and smooth account transition is sought by sales management? Is the salesperson working smart or is he setting himself up for a loss after a long sales cycle? How is he or she able to close business without using most of the available resources? What resources are available that are not being used? Sales management must ask all of these questions and make sure the answers are reasonable. Otherwise the salesperson is probably wasting irreplaceable time and resources.

Teamwork is a term which is greatly overused. This term used to mean everyone pulling together in the same direction toward a commonly understood goal. Today, I rarely see true teamwork in my travels. Usually management has not clearly communicated the goal(s) so all can see and understand how their job directly relates to the success of the overall organization. As a result, many individuals seem to be working their own agendas, which is to do just enough to keep the boss off their backs. Their personal agendas seem to have become more important than the company's success.

Too often employees seem to be working at their own pace. "What's the rush? Sales has waited this long, they can wait a bit more," seems to be a common adage. Wrong! "They" are not sales; they are the customers. These same customers can easily buy from another company. All employees can and should contribute in a competitive sales situation.

Teamwork does not spontaneously occur when individuals gather to discuss an opportunity. Teamwork occurs when all the individuals around the table understand the goal, accept their role in the process to achieve that goal, and truly commit to winning.

Team goals are important, but not at the expense of individual accountability. Individual initiative and hard work seem to get overlooked by management when teams are formed. Team measurements must be utilized constructively and not as a shield from accountability.

Often when a key sale is lost we remember how hard everyone on the team worked. It is a sad loss; we cannot punish everybody. We cannot single out one person for not doing their part. "Oh well, let us move on to the next opportunity" is the prevailing attitude. However, without constructive feedback what keeps individuals on that team from making the same mistakes over and over? Alternately, when a big deal is won, it is interesting how everyone all of a sudden claims to be a part of the "team" – the team that won.

Do team measurements reward everyone and especially the real heroes after a team victory? Or are the rewards parceled out to ensure everyone is treated exactly the same, no matter each person's contribution? Invariably there is someone viewed by other members of the team as a non-contributor. Is this perspective true? When the group attains their team goals, does everyone partake in the reward? When the team does not achieve its goals is everyone admonished fairly? These trade-offs must be balanced to ensure all members of the team feel fairly valued and rewarded!

The fundamental basis to engender an environment of teamwork is respect – respect for human beings, diversity, other people's jobs, and recognizing all people's need to feel worthiness and a part of the success of the organization.

It is of vital importance that the sales and marketing organization set an example for the entire company. Are sales and marketing people polite, courteous, and appreciative in all of their dealings with other employees? Are all employees encouraged to be a part of the selling process? Are non-sales employees recognized and rewarded consistently when sales victories occur? Rewards can come in many forms:

- plaque
- stock option
- small check
- large check
- lunch
- pay raise
- dinner
- three-day cruise
- sincere round of applause
- personalized letter from the president with the employee's name spelled correctly

Each salesperson has the responsibility of ensuring every person who supported their march to victory is remembered and recognized. Sales management while struggling to stay in the daily race do not always see all members of the pit crew. An environment of sincere respect and appreciation will foster more teamwork than any speech or memorandum from a weenie.

* * *

Sales: The Core of the Selling Machine

Different territories and accounts need different emphases when it comes to sales skills and level of expertise. To successfully sell in a competitive environment, aligning appropriate skills and experience levels with the needs of the company's accounts is critical.

Sales skills, in this context, relate to the inherent abilities of an individual to identify and close business. Early in my sales management career, I made an interesting observation about the different types of sales skills required in a competitive sales situation. I had just accepted a new assignment to form a sales branch to pursue federal and commercial business in Washington, D.C. I initially hired three salespeople to get the ball rolling. As the sales team began to form, I noticed a phenomenon which was amazing. One woman who worked for me had the uncanny ability to find people with money who needed our products. She was the best *tracker* I have ever seen. She did not seem aggressive in closing the business but she could consistently find potential customers. Another woman who worked for me was an aggressive closer. Once the business was found, she was expert at closing the deal – getting a signed purchase order. I began to think of her as my *killer.* Finally, a man who worked for me did not like finding new business, but he was a master at getting additional revenue from an existing account. He was clearly a *skinner.*

I actually had three salespeople working for me who displayed these sales skills quite separately. Few people are adept in all three areas, however, a consistently successful salesperson needs all these traits among their sales skills.

Do you truly understand the strengths and weaknesses of your sales team? Do you know what your salespeople's interests or capabilities are in these three areas? Do not assume all your salespeople are proficient in all three areas! I, for example, like "tracking and killing" but avoid "skinning." For maximum impact, ensure the sales skills of the individuals within the organization are mapped to the needs of the customer accounts.

Additionally, ensure your best sales talent is calling on your company's most critical accounts. It amazes and angers me the number of times I have seen large, critical accounts covered by a junior salesperson. This is ludicrous! The most senior salespeople should be assigned to your most critical accounts. Sales management 201! A key customer's potential revenue is an expensive way to train a junior salesperson. You can bet the competition does not have a junior salesperson calling on your key account. Sales management will deny this happens often. How often can your company afford to have this happen at all?

Let the junior salespeople call on new accounts or unexplored territory. Let us find out up front who is meant to be in sales. Yes, the junior salespeople need training, coaching, and mentoring. But do not do this development at the expense of the stockholders and other employees' jobs.

Can you Help me?

A senior salesperson stopped by my desk and asked me for assistance. This person had about seven years sales experience and had never failed to exceed his assigned quota. He had received about every sales award available from the company. The salesperson said, "Another computer company is getting into my account. I do not know them well. I have not competed for business in my account. I am not sure what to do. Will you help me?"

I was dumbfounded. I appreciated the salesperson's candor and willingness to ask for help. But I could barely answer the person, because I found the statement to be incredulous. This was not a senior salesperson! This was a senior order taker.

Are your "senior" salespeople really senior? Are they ready to compete for business? Do they even know a competitor is penetrating their key account? Are you relying upon these people to target new accounts for penetration this year? Are these salespeople being promoted into sales management in your company?

When was the last time sales management took a hard look to make sure their best sales talent is in fact calling on the company's most critical accounts? One company I worked for seemed to have mass account reassignment every summer. Of course this company also had a reward system in place which encouraged account hopping by its salespeople. Loyalties and trust with the customer cannot be built over time when the customer sees a new set of faces every twelve months. Alternately, accounts are not fiefs of the salespeople assigned for eternity. The sales manager has a fiduciary obligation to ensure key accounts are getting the best sales coverage possible. This is a difficult charter. Sales management must balance the need to build solid relationships with customers with the requirement to ensure each account is being called on in a professional and aggressive manner.

* * *

Sales Support Analysts

Early in my sales career, I became spoiled. I had the opportunity to work with Chuck. I did not know Chuck well at the time, but everyone said Chuck was a guru – he had been around the industry a long time.

Chuck quickly explained to me his role as a sales support analyst: His job was to work hard to help me achieve and exceed my quota. Period! Ah yes, I became spoiled.

Chuck had a rare work ethic. He worked past 5:00 P.M.; he came in on Saturdays; no customer's question was stupid; he liked to go on sales calls with a salesperson; he liked to give presentations and was prepared to do so on short notice; and he actually liked salespeople. Can you imagine that? He knew his job and he knew what salespeople were supposed to be doing to be successful. No task or request was beneath Chuck, if getting it done would help close business.

Sales support analysts have a tough job. They must

maintain technical competency on a range of subjects or their knowledge has little value in two years. They have to deal with irate customers, aggressive competitors, impatient salespeople, little management support, lack of resources and quality training, and juggle product and program business demands each week. Not many people have the fortitude or dedication to survive in this frenetic work environment.

A sales support analyst is a valuable resource. They must be managed on one hand and left alone to do their jobs on the other hand. They are not independent consultants who flit in and out of sales situations, nor are they resources to be consumed for seventy hours a week, months on end, for example, performing a complex customer benchmark.

It has been many years since I worked with Chuck. I can count on the fingers of one hand the sales support analysts I have worked with who truly understood and accept their relationship to the sales function. Their specialized skills and knowledge, when appropriately applied, comprise a critical component of the total sales team.

There needs to be a close working relationship between sales and sales support – not arm's length, not casual, but one of trust and cooperation. I have met sales support people who act like independent consultants and believe their paychecks come from Payroll. I have also met salespeople who treat their support team as "gophers" and never share account or selling strategy information with the full sales team. They do not see their support analysts as valued members of the account sales team. These extremes are intolerable in a hi-tech sales environment.

To have mutual accountability, related measurements and rewards, and an almost equal voice during account plan development both sales and sales support must report to sales management. Customer services must also participate in the winning-new-business strategy sessions but are seldom included in most companies. Their understanding of customer environments is invaluable. The days of talking teamwork are over! You either cooperate and execute as a team or you can keep going to account loss reviews. Sales and sales support are interdependent and crucial ingredients for a well-executing selling machine.

* * *

Sales Support Career Path

For growth and advancement, sales support analysts have a variety of options: a technical track, a sales option, or a sales support management position. None of these options are always right or wrong; however, it is important to understand the individual's strengths and weaknesses as related to each option. Too often advancement can occur based on individual desire or immediate corporate need, rather than appropriate matching of skills to the longer view.

The technical track has three levels which allow the individual to progressively increase his or her skills and value to the corporation. Given the exploding growth and increasing complexity of technology, this path offers continuous challenge and opportunity for strong technical support analysts.

The first track is a "Mr. or Ms. Fixit" kind of person. He or she can answer routine product questions, get the right technical manual when needed, and understand the inputs and outputs of the component. These analysts work and think at the product level. Most of these analysts work in the post-sales environment where they address routine customer questions and problems.

The second kind of analyst works at the system level. He or she appreciate how the product is used in different application environments and the features and benefits of product usage. They understand the design attributes of the product and even have their own ideas on how the product could be made better.

The third level is the systems architect or visionary who can articulate and write about how a given productor system will help the customer better achieve their organizational mission. This person usually has in-depth knowledge in more than one discipline, for example, database and network management.

Another career option for sales support analysts is a move into sales. Some companies view this transition as logical. Some sales support people believe this will be a nice change of scenery. Over the years many companies have lost competent sales support people and gained a subpar salesperson because of these misconceptions. A level of technical knowledge is an asset to a sales person, however that is not the critical component for success. The ability to aggressively compete in the marketplace, develop mutually rewarding relationships with customers, and define and implement winning strategies are key to success for a sales person. The job is not all fancy lunches and commission checks.

Very few of these salespeople I have met had the courage to return to their sales support role after a poor showing in sales. Very few were given a chance to return to a job they were capable of doing well. What criteria does your organization employ to ensure this job transition is effective and successful for all parties? Is there an option for a graceful return?

The Demotion

Bruce shuffled into my office and sat down quietly. He was a polite, soft-spoken salesperson. Bruce knew I was unhappy with his performance the past few months. We had talked repeatedly. He was 25% over quota but something was wrong. I had thought about this meeting for days; it was time to take action. I was going to have to demote Bruce.

After a period of brief discussion stating the purpose of the meeting I said, "Do you understand why I am demoting you?"

"Not really. I am over quota and get along with everyone in the office. Have you had any complaints from our customers?"

"No, I have not. Bruce, you are a nice person and are doing the job to the best of your ability. You are over quota and will earn $150,000 this year. But there is one problem – our competitors have won the last two large deals in your primary account. You have not won any competitive business in the last year. You are not paid by this company to process orders and take a customer to lunch twice a month. You are paid well to keep the competition out of our knickers. You have totally failed to do this. I have offered my assistance and other resources to no avail.

Do you have anything new to say?"

"No."

"Then, as of March 1, you are being returned to your former position in the company. You did a good job there and can still make a valuable contribution. I will get with you Friday to begin account transition."

Bruce left the office quietly. We spoke little the next couple of days. He resigned on Friday and went to work for a direct competitor as a senior salesperson. About two years later he gave up his hi-tech sales career and went into real estate property management where he has been quite successful.

Finally, other sales support analysts decide it is time to become – a sales support manager. You know, more money and less demands....

The sales support manager is an important position in the world of hi-tech sales. The technical skills that may have spelled success as an analyst are not the only skills needed to manage the technical support team. The analyst support manager must understand where each of his or her team members are in terms of their skill levels – not only technical skills, but also sales and business skills. Then, he or she must prioritize workloads for his or her subordinates, thus minimizing the frustration caused by conflicting demands, track the requests for help and assistance, forecast support trends concerning new technologies. The sales support manager must also ensure staff is getting training and up-to-date information on new technology, and guard against analyst burn-out that is the inevitable result of the continual demands on their time. Sales support management requires strong management skills, not just technical ability. Who are you promoting into sales support management — and why?

Chuck, my sales support team partner, did not want to be a sales representative or a manager. Chuck understood his job and the roles of the others on the sales team. Chuck wanted to be a technical guru in a sales support role. Can each member of your sales support staff clearly articulate their role and path for the future? Does management know and agree with that path?

Sales support analysts are critical to the success of the sales operation. Understanding the skills required to perform the tasks and managing the expectations for growth and advancement are crucial to the long term professional development of the total sales organization.

* * *

Sales Support Management

The sales support manager can be a linchpin to the sales operation. Competent sales support managers are worth their weight in gold. Appropriate and proactive management of sales support resources provides the organization a very powerful tool in the pursuit of competitive business.

The sales support manager must understand the technical, sales, and business skill levels for each of his or her team members. Proactive management requires an understanding of and ability to influence organizational behavior and group dynamics. Who likes to work with salespeople? Who likes to give (focused) product presentations? Who will take (and is ready to take) ownership of a critical account's technical problem? Who is comfortable knowing one subject well and resists training outside their favorite technical subject? Who is working thirty-five hours a week and who is working seventy hours a week? Who is organized and does not overcommit? These questions must be asked and understood by the sales support manager.

Management also requires plain old organizational skills. Understanding, documenting, tracking, and communicating the skills of individual staff members can be complex. There are two key matrices which can help make this magic happen.

The Analyst Skills Matrix provides an example of a tactical management tool used to track and identify team strengths and weaknesses. In a snapshot, the sales support manager can evaluate team-wide talent for coverage and exposure. The selected skills should reflect, in priority order, the critical applications and technologies needed by the organization. All analysts should be listed and ranked for each item. This simple tool can be used to help develop individual training and development plans to shore up the weaknesses of the team.

For example, several observations can be made by examining *Exhibit C*. Networking seems to be covered pretty well by Dawn and George. The lack of Internet expertise sticks out like a sore thumb. Charles and Dawn appear to be the workhorses. Bill and Janet do not seem to be key contributors. If I need an expert-level presentation for Security or Internet, I am in trouble.

Be proactive! Are you getting the most you can from your sales support staff? Have you identified the specific technical development and training needs of each person? How often do you update and review this information? Do it now!

Analyst Skills Matrix

Analyst	Data-base	Security	Net-working	EDI	Imaging	Internet
Bill	2	2	1	0	0	0
Charles	3	0	2	0	4	0
Ruth	0	3	0	0	0	2
Dawn	3	0	4	4	0	0
Frank	0	0	0	3	2	2
George	1	0	4	0	2	0
Rajan	0	0	1	0	0	0
Maryam	4	1	1	0	1	0
Janet	0	0	1	1	1	0

4 - Expert. Years of applications experience. Can handle almost any question. Can present this subject at any level desired by the customer or salespeople.

3 - Some experience. Can handle most questions. Can give a product presentation competently.

2 - No experience, but has some training and interest in this subject. Is not ready to give subject matter presentations.

1 - No experience nor training in this area.

0 - Not applicable to this person.

Exhibit C

The Product/Program Matrix shown in *Exhibit D* provides the sales support manager another tool to assist in juggling product and program business. The Matrix provides a clear picture of who is over- or underworked. It helps keep the team focused on sales management's business priorities and (past) commitments. This tool should help bring order into an otherwise chaotic environment. *Exhibit D* should be updated at least every two weeks to ensure resources continue to be focused on critical business priorities. You can also use *Exhibit D* to assign the more junior support staff to product or program opportunities under the tutelage of a senior analyst. This mentoring role is important to all parties. Stop betting your success on the same workhorses!

Most sales support managers I have worked with spent their time studying requests for sales support and qualifying new account sales opportunities more than sales management. Instead of managing a logical process, they were continually reacting to a constantly changing environment. Instead of understanding the strengths and weaknesses of their support team, they avoided prioritizing their workloads, hoping their people did not get too frustrated and burned out. I have only worked at one company where there was an attempt to manage this valuable resource in a proactive manner which seemed to be effective. Their management approach resulted in careful usage of their analyst support staff's time. A balance was struck between installed and new business, scheduled technical training and development, and ever changing sales management priorities. Does this sound familiar? Until corporations understand how to proactively manage this key resource, the sales support function will remain a problem area for the sales organization.

Product/Program Matrix

Key Activities for October

Analyst	Product Opportunities	Hours/ Week	Programs	Hours/ Month
Charles	DEF Corp.-Imaging pitch	8	ABC, Amson Corp.	80
	GHI, Inc.-Networking demo	8		
Ruth	MNO Assoc.-Security white-paper	12	JKL Bureau	20
	PQR-Internet questionnaire	8		
Dawn	STU Ltd.-EDI benchmark	20	Env. Science Div. Inc	30
	GHI, Inc. -Networking demo	8		

Exhibit D

Consultants

When I was new to sales, other "senior" salespeople told me to ignore consultants. It was explained that consultants are people who want to meet with you, seek free training, documentation, and product demonstrations, and never buy anything. Given that advice, I avoided consultants for years. After all, given limited resources, why should a good sales person waste time and resources on a group of people who never buy anything? Because they are an important and valuable sales channel to the marketplace!

Consultants play a vital role in conveying their perceptions of industry products and services to a prospective customer or marketplace. The majority of my customers, when I was pursuing large sales opportunities, retained consultants to assist them in the decision-making process. I do not avoid and ignore consultants any longer. Consultants are a valuable selling resource and should be managed as such.

An information technology company can develop a simplistic, low-cost consultant program. For example, every four months you could have a Consultant's Day. Consultants are invited to a presentation with a strategic marketing message (real meat), given a professional product demonstration, a full set of product literature, coffee and pastries, a management point of contact for general questions, and a sense of feeling important to your company's future success. This half-day consultant's program can go a long way toward establishing your company firmly in a specific market segment. Generally, most consultants specialize in specific applications, geographical areas, or industries. You should be able to identify and begin relationship building with a specific consultant community. This low-cost investment can pay off in many ways.

Consultants sell knowledge or advice to their clients They especially like to sell current knowledge or mark insight. Therefore, they need to understand the newest technology and concepts. Hence, they will call you for free documentation, product demonstrations, and so on Give it to them. Do not slough them off to some junior marketing coordinator. Do not ignore consultants; treat them as you treat any key potential customer!

* * *

Resource Allocation

The appropriate selection and application of scarce resources is a constant challenge in any competitive sales situation. There are a variety of tools in the salesperson's bag, but selecting the right tool at the right time for the right effect can make the sale happen. Constant negotiation, renegotiation, and cajoling for resources to get help to close a new account does not build a winning team – does not create synergy. Alternately, using resources without a plan or strategy results in waste of irreplaceable time and assets.

To ensure all resources are being considered and appropriately applied to a competitive sales situation, a Resource Allocation Matrix (RAM) is a valuable tool in account management. *Exhibit E* on the following page provides an example of a RAM. The RAM allows sales and sales management to focus all appropriate resources on an account.

A RAM can be used to plan and track application of resources within an account in a consistent fashion. This tool helps the sales organization identify where specific resource requirements should be applied to key decision makers in the account. A check is entered into the appropriate cell when an activity is defined as required; a date is entered in the cell when the activity is completed. Regular review and updating of the RAM between sales and management can give quick insight into account planning and sales activity. For example, you can see if the services manager has been to the account or if senior management has called on the Chief Information Officer (CIO). Customers definitely like and expect to see more faces from their strategic suppliers than just the salesperson's.

Customers also like to receive good solid technical information to assist in the decision-making process. The sales support staff is a critical resource in the art of selling. The ratio of sales to sales support has been in the 1.5 to 4 salespeople to each sales support person where I have worked. The more complex and expensive the product and services, the lower the ratio. The longer the sales cycle, the lower the ratio. I have actually had customers ask me for this ratio to help them ascertain if my company would be able to deliver timely support.

Resources are limited in every organization. Too often resources are applied when available, not when required. Developing and application of a RAM helps the organization apply resources in a planned fashion that takes full competitive advantage of the resource. In the world of limited resources, no organization can afford to expend resources inefficiently.

Resource Allocation Matrix (RAM)

Decision Makers and Key Customer Names:	President	CIO	Engineer.	Purchasing
My Company's Resources:	Banwell	Wilmer	Aston	Gorman
President				
VP of Marketing				
VP of Sales				
Dir. of Engineering				
Tech. Documentation				
Training Courses				
Sales Manager				
Service Manager				
Sales Support Mgr.				
Mgt. Presentation				
Tech. Presentation				
Breakfast				
Lunch				
Seminar				
Custom Demonstration				
Reference Accounts				
HQ Visit				
Benchmark				
Letter of Commitment				

Exhibit E

Diversity

A well executing selling machine's parts consist of more than simply the functions of individuals. In business, as in life, diversity lends richness to the process. The topic of diversity covers many areas: age, gender, race, ethnic background, life style choices, and so on. Respect for diversity and the value of individuals is critical to a successful business today. However, diversity in the work force has become a complex and controversial subject.

For example, I was once asked by a senior executive in a large, prestigious account to replace a black woman who was working with me on a large business opportunity worth over $20 million. He stated that he was concerned about her "ability." I had two private conversations with this executive to try to understand his concern about her ability. The rationale for his request remained unclear to me. I insisted this woman stay on the team. I told him her skills and experience were needed. The executive backed off and the matter did not arise again. I never told her about this conversation.

Alternately, on another occasion I asked a black sales manager to join me on a sales call to visit a minority-owned firm in Washington, D.C. I explained to this manager his presence would help our image as a company and strengthen our position in the account. The black sales manager was offended by my request and refused to go. My objective was to increase our chances of winning this deal – no more, no less. I followed through alone; we did not win the business. Was I wrong? Did I go too far to try to win? Was I insensitive to the black manager's feelings?

We know there are differences between people. Recognizing those differences and understanding their impact has a political, social, and business impact. To value these differences is appropriate. However, is it incorrect to recognize that an individual's sex, color, age, or other characteristic can affect a business situation? This is certainly a double edged sword! The effect can be positive or negative.

For example, based upon my experiences, I believe there are differences in the expectations of men and women in similar sales positions. Until the late-eighties, I had seldom worked with, and never had worked for, a female. Up to that point in time, I had been selling primarily to OEMs and VARS in technical application environments. I do not recall meeting with a female engineer until the mid-eighties.

Two female salespeople worked for me in the 1980s. They worked equally as hard as the men, but for some reason, seemed to be able to get the door opened more quickly into a new account. Furthermore, they seemed to get a lot more information on the first call. It was as if their customers wanted to talk with them. I believe this is a natural advantage particularly when a saleswoman is calling on a male. I believe most male customers will meet quicker with a female salesperson, rather than a male salesperson, all things equal. I am not saying there is any lecherousness; it is simply male curiosity. Who is the female voice on the telephone? Does she really know her stuff? However, once that door is opened, it can be harder for the woman salesperson to break down the barriers created by a predominately male enclave.

Pub Crawling

I found myself in an especially competitive sales situation competing with an intelligent and attractive saleswoman, Sandra. She not only worked hard, she also had an extensive technical background. Sandra had a great personality and is the kind of person you want for a friend. More importantly Sandra had access to the customer's senior management. This was one of her key advantages.

During this highly competitive sales opportunity, I had trouble getting access to the key technical decision makers. I knew I had a better product, better price, and it was my job to make sure the customer knew it and, of course, selected our computer systems for their environment. It frequently seemed the decision makers were usually either gone or meeting with Sandra. I had to get access to them and an understanding of their decision making process for the computer system selection. I discovered two of the key managers spent one or two evenings a week at a local pub, having drinks and shooting pool until the wee hours. I immediately started going to this pub two nights a week. I would hang around waiting for them. They would arrive and while always polite, were tight-lipped. They knew why I was there.

I sipped beer with them, shot some pool, ate some barbecue ribs and discussed football, baseball, and so on. I did this steadily for five months. I never asked direct questions and seemed laid back if the sales opportunity was mentioned. Slowly the walls came

> down and their technical concerns about getting the job done with the least risk began to surface. Based upon technical concerns I heard them mention, I then put together a technical solution which got their attention. Eventually I won the deal – the initial order was for $4.5 million. This victory was the biggest win of my career.

Would most female salespersons have stayed until the late hours twice a week "drinking with customers?" Some might say it would not have "looked right." This was my advantage as a male, so to speak. Women today do entertain customers, play golf and racquetball, for example. But do they have full access to male customers? I am not sure.

When I was a hiring manager I tried to match the style and personality of a salesperson to an available sales territory. I did not care whether a person was female or male, young or old, or black or white. I had accounts who liked female salespeople, or wanted a younger guy to play racquetball, or wanted to see a black salesperson, or wanted an older person who had been around. For the sake of establishing relationships with customers, achieving sales goals, and keeping paychecks flowing throughout the company, I tried to hire people to meet the personality and temperament of my customers. This approach is subjective, but my goal was to put the right person in the right job at the right place to increase our probability of success. With rare exception, my approach worked over the years.

One company I worked for took diversity in the workforce to a fine point. There were going to be female and male, black and white (and other ethnic groups), old and young, technical and non-technically oriented salespeople in the sales branch no matter what! People were placed into sales branches to ensure the "right mix" was achieved with no regard to customers' temperaments or expectations. Obviously the branch sales managers tried to balance and match personalities as well as possible, but they had no authority to reassign people when potential concerns arose. The company lost business on occasion because diversity in the workforce was sacrosanct and not subject to discussion. This company still has this policy in effect, to my knowledge, even as it continues to layoff employees.

Welcome to the Account!

My manager called me into his office and said he was assigning me to a long-time account of the company's. This account was located about five hours by automobile (each way) from the office and past sales coverage had been spotty. Air travel to this account was not practical due to its somewhat isolated location. The salesperson covering the account had to be willing to go there at least twice a month. I told my manager I would handle the account and asked him who I should contact first. He told me to go see John, an engineer in the account.

I soon scheduled two days to make my first visit to see this customer. I called John and introduced myself. He told me to call him during the afternoon once I arrived in town. After arriving around 3:30 P.M., I called John to ask him if I could come by his office that afternoon. John told me he could not see me until later, and for me to go to a nearby liquor store, buy one ice-cold case of Genesee Cream Ale, and meet him at 5:30 P.M. He gave me specific directions to a park one mile from his office, to the second parking lot on the left which had three or four picnic tables. I said okay even though this sounded peculiar.

At 5:30 P.M. John arrived at the park, said hello, shook my hand, then opened two of the beers, one for me and one for him. We began to chat. He wanted to know my background, where I went to college, where I was from, whom I had worked for, what I liked and disliked about my job, who I knew in the D.C. area, who I knew in his company, if I had any technical experience, and on and on. We continued to sip the beer and talk. Gradually the conversation shifted and John began to tell me about himself, his family, job duties, hobbies, education, and so on. The conversation continued until 8:30 P.M. at which time we were drinking the last of the beer and I was becoming quite tired.

Suddenly after his last swallow of beer, John said, "I'm hungry. Let's go get some food. We need to talk business." We got into my car and went to an Italian restaurant in town, where we had a wonderful meal with two bottles of red wine. We discussed a new project he was planning. He became a valuable inside salesperson for me over the years.

I found out later this was John's standard welcome-the-new-salesguy-to-the-account. Given today's sensitivities and concern for political correctness, I wonder, does he still invite the new salesperson to share a case of beer in the park? And if the new salesperson is a woman, would this behavior be considered acceptable?

Davis' Rule Five

Drive the sales process before it drives you.

In the Beginning

A sales career is not something I planned. I never thought about being a salesperson. While millions of people work in sales positions, it is not a career most people seek. It often starts out as a job you do until you find something "better." Of course, better may not come along. While in high school, I thought about college, the draft, and getting a "good job." When I was eighteen years old, a good job meant having job security and making more money than Dad. Most people today would consider a good job as one offering job security and a higher than average income.

In 1975 I worked as a junior FORTRAN programmer for Tesdata Systems Corporation in McLean, Virginia, responsible for installing a simulation software package onto large IBM and Unisys computer systems. One morning my manager explained to me that sales of the software packages were very slow, as a result, my installation services were no long required. Hence, I was laid off.

I assumed my career would continue in a technical direction. However, after two weeks of financial independence, I decided I needed to find a job. I grabbed my resume and the yellow pages of the telephone book. Beginning with the A's, I looked up a company, drove to their office, and upon entering, asked if they were hiring. I was almost hired by Data 100 Corporation. However, when that seemed unlikely, I drove to Data General's offices in Falls Church, Virginia. After my initial question, "Are you hiring?" I was whisked into Tom's office, who asked me if I wanted to be in sales. I said "Absolutely...!" and thereby stepped into the next phase of my career.

Tom informed me I would be calling on the civilian agencies of the federal government. While I knew the government had certain "rules" about buying things, I figured it could not be all that complicated! Sales was just talking with people, answering questions, providing a brochure, and writing an order. How complicated could it be?

I asked my boss which specific customers might buy our product – a 16-bit mini-computer, since the product had limited software, that is, a FORTRAN compiler and a manual? He told me, "Call on the National Aeronautical and Space Administration (NASA)." I asked him who in the company knew someone at NASA. He told me, "Go see Jack." He explained Jack was a senior salesperson who used to call on NASA. So, I immediately proceeded down the hall to see Jack. I stuck my head in the doorway and caught Jack in the middle of sorting manila file folders on his desk. I politely introduced myself and asked him for his contacts at NASA. Jack simply stared at me and slowly handed me a folder from the middle of a stack on his desk. He did not say a word; he just stared at me. I thanked him and returned to my new office eagerly looking through the file for a name of someone ready to buy.

I later understood why Jack gave me that peculiar look. Jack, when I interrupted him, had just been laid off, and was clearing out his desk. I had just blandly asked him to hand over the results of his years of sales effort at NASA. Jack and I stayed in touch for many years and I later thanked him for his politeness.

While scanning Jack's NASA account file, I came

across a name underlined and boxed in yellow. This must mean something! I immediately dialed the telephone number to schedule an appointment. A secretary politely scheduled the appointment two weeks later. Finally the day arrived and I grabbed technical brochures and headed to NASA looking for a quick sale.

First Sales Call

Here I was, at 1:30 P.M., sitting in this large office, with a customer who had Chief in his title. This was it! Sales was not so hard after all. I immediately began to talk about our newest minicomputer using a chalkboard to diagram our computer system's buss and why it was really the fastest in the marketplace. After fifteen minutes, I noticed the customer's eyelids beginning to droop, his head nodded forward, his shoulders sagged, and finally the customer was sleeping soundly with intermittent snoring. I called his name twice to no avail. I thought about shaking him violently, but I figured that on the first sales call that might not be a good idea.

I slowly got up, went out of his office and told his secretary he was asleep. She looked at me briefly, then got up to see for herself. When she came out of his office, she said, "You're right," and sat down at her desk and went back to doing her work. I surmised this situation had apparently happened before.

I walked back into his office quietly and left my business card and brochure on his desk. I left his office slowly, making sure I did not disturb his slumber. Obviously my presentations were going to have to become more interesting! Maybe this sales job was going to be a bit more complicated than I had figured....

Sales Is an Art and a Science

Russian chess champion Sergei Tartakower once said of a chessboard that was set up for a game, "The mistakes are all there, waiting to be made." So it is with a sales opportunity. Sales is both an art and a science.

Selling requires developing a business relationship between two divergent organizations. Customers today are not buying products or services, so to speak; they are buying a consistent, reliable response to their anticipated and unanticipated business requirements. This requires not only an understanding of their business, but more importantly a commitment to their success. Your dedication to your customer's success will lead to mutual success.

Sales is a logical process. It is more deterministic in nature than many business professionals realize. Theoretically, resources are applied in a consistent process and a result is achieved. Examples of resources are technical staff, customer references, hardware, software, documentation, training, and management. (Please refer to the Resource Allocation Matrix in Rule 4.) A salesperson should be able to tap into his or her company's resource pool and apply the appropriate resource at the right time, in the right amount to the selling process.

Understanding and application of this consistent selling process will result in a positive business relationship which should, in turn, produce orders. Sound pretty easy, huh? If you are reading this book, clearly you know it's not as easy as it sounds.

Sales is where the rubber meets the road. Sales must be the conductor for a well-rehearsed orchestra. The customer must hear music and see a seamless organization with one consistent focus – customer satisfaction. If the sales function is not performed well with seasoned professionals then the company will not win its fair share of competitive business. The company will just continue to struggle to stay in the race.

* * *

Who's in Charge Here Anyway?

Who owns the business or revenue, so to speak, in a company? When each employee is asked who is responsible for revenue, how would they answer in your company? Every organization is comprised of people with different backgrounds and perspectives concerning the company's mission. Furthermore, the traditional functions of business such as manufacturing, finance, engineering, sales, marketing, and services are not uniformly strong across many companies in the hi-tech world. Each company seems to have a couple of functions which seem to lead or "steer" the company and dominate the strategic decision-making process. For example, one company seems to be led by engineering, while another company appears to be "run" by marketing. But in terms of revenue, who is ultimately responsible?

Many people may assume the answer would be sales. I have asked this question during employment interviews at different firms and received a variety of answers including: engineering, marketing, sales, services, and even "that is a matrixed responsibility." This is a critical matter, because if sales is not perceived as being responsible, accountable, and vested with the authority and resources to secure revenue, then the company is in trouble.

Imagine a U.S. Marine captain on a ship looking through binoculars at the enemy entrenched on a hill overlooking a sandy beach about a mile from his or her ship. He is responsible for taking "hills." This is why his job exists. There are resources in place to help him take hills. He knows if he does not take some number of hills each year, he is not considered as having fulfilled his mission or achieved results, so to speak. He also knows he cannot invest too heavily to take one hill – some of his assets and resources are not replaceable.

Now, the time has come for him to take this particular hill. Time for action is at hand. What is the captain thinking? Is he thinking about the hill itself, the enemy inside, the dangers of a frontal assault, and so on? NO. If he is a professional, he is thinking about step one – getting his people and material safely off the ship into the landing craft. Then he thinks about step two – getting through choppy waters and avoiding mortar shells on the way to the beach, arriving with all his resources intact. Then, he thinks about step three – getting out of the landing craft, across the beach, and behind a sea wall without taking losses. Then, he thinks about step four – getting from behind the sea wall, up the embankment to be in a proper position to be able to successfully take the hill. Then, he thinks about step five – actually taking the hill in an aggressive, but prudent manner.

So it is with hi-tech sales. Many times management will say "close the deal," or the classic, "Do not come back without the order." But all too often neither management nor the sales team have done all the intermediate steps to properly position themselves to get the order. If you do the intermediate steps right, you will be in a position to successfully capture the business. If you do not do the intermediate steps right, then you will not capture the business and will sustain losses. Sales is a process. As managers, do we and our salespeople know what these intermediate steps are and their correct order? We should; that is our responsibility.

Let us imagine further, you are the same Captain charged by the General to take hills. You know you will need artillery, air power, and maybe flame-throwers to help your people get close to the objective without too much exposure to danger. Once you have formulated a plan to approach and capture the objective, you approach your organization's support elements and are told: "We are busy today." "I do not think that is the way to do it." "I do not like your plan." "What is in it for me?" "I will get back to you in two weeks." "I am not sure I can help you." or "The lieutenant is not here." This is the exact situation senior management in large companies often places the sales organization. Somewhere the General forgot to tell the rest of the army they are there to help sales take "hills." It is absurd to make an organization accountable to get a job done and provide no real mechanism to secure the cooperation and commitment of necessary support resources.

Is there a significant breach between salespeople and the support organization? A salesperson needs the unique skills and knowledge of the support organization to help address customer concerns and issues. However, if a support organization is unresponsive to a salesperson's requests, the ability to professionally respond to customer issues becomes exacerbated. How are these so-called customer service and account support representatives really measured and rewarded?

Why do some employees suddenly become hard of hearing when the sales organization requests support? Is it:

- a lack of understanding concerning the proper role of sales to the success of the organization and in turn their own department's relationship to the sales function?
- ineffective measurement, reward, and recognition programs from sales and higher management?
- an insensitive or arrogant attitude by the salespeople toward the support staff?
- an environment where the term "teamwork" has no real substance except when a Senior Vice President is visiting the office?
- an environment where management has pampered and coddled the salespeople to the detriment of the rest of the organization?
- functions in the field, reporting separately back to headquarters and therefore having no success linkage to the sales organization at a local level?

All of these questions must be asked and discussed in a candid fashion for real progress to be made, and for a positive relationship to evolve. Catherine, a sales support person in a chemical company, told me all of the sales-people were commonly referred to as "the assholes." This comment may sound humorous, or be heard often, but it is deadly to the success of an organization. Top management, sales management, marketing, and the salespeople themselves must constantly address sales' image and leadership role within their organization. Disrespect and dissension between sales and other parts of the company is unforgivable. It produces incredible drag on the potential success of the company. It is a luxury tax no one can afford.

If sales does not have authority and "vesting" by other functions in the company, then the company cannot design a selling model supported by all company functions. There must be a single focal point of accountability. For example, four managers in four departments cannot all be held accountable for closing a new account or increasing revenues by fifteen percent this year.

The lack of commitment to be a focused selling organization causes an insidious drain on efficiency and profits. It causes resources and departments to misdirect energy and lose sight of the importance of their (sales) support roles. Functions, acting in good faith, keep going in their own direction instead of getting 100% behind who should be the trailblazers – sales.

There must be a focal point of leadership. When there are separate organizations with separate reporting structures, separate priorities, and unrelated performance appraisal goals, why would anyone outside of sales take ownership to help sales take "hills"? Resources in the field should work for sales with rare exception.

* * *

Empowering the Selling Machine

In this age of efficiency at all costs, "right sizing," and re-engineering business processes, has your company focused on empowering its selling machine and gone even further to take the "shackles" off its sales organization? Have you really vested the sales organization with the authority to get the job done?

Many companies have placed "handcuffs" on their sales organization and don't even know it. These handcuffs serve to constrain their authority and ability to respond to customer problems or competitor's actions. For example, Operations, Finance, Manufacturing, and Marketing believe they have a legitimate stake in the day-to-day decisions and actions of sales management. Maybe they do; maybe they do not. If sales management is unskilled or unable to do business negotiating, set discount levels, or reallocate scarce resources, for example, then you have a much bigger problem than you realize. What are sales managers being paid to do? What should they be doing? When does senior management realize that sales management is not getting the job done? When market share is declining or a smaller company keeps winning competitive deals? Then it is too late.

Senior management or sales management may have policies and procedures intended to support the sales process, for example, sign-offs or business reviews. The fact is these policies and procedures can hamper the sales process if they have a primarily internal focus. Clearly, communication and information sharing is important across organizations within the corporation, however they consume time which could be spent in face-to face meetings with customers. Balance is the key. The following signposts indicate that an internal focus has become more important than an external or customer focus:

- Salespeople are in the office more than 50% of the time each week.
- More hours per month are spent in internal meetings than in face-to face meetings with customers.
- Salespeople are regularly and suddenly asked for data beyond an expense report, forecast, and account sales plan.
- Multiple sign-offs are required before a customer-related decision can be reached.
- A meeting with middle management is required before a low-level decision can be made.
- All issues must be in writing before a decision can be reached.
- Management avoids speaking with the customer until a consensus is reached.
- Employees are writing their own performance appraisals because their managers are unfamiliar with their actual job assignments.

These signposts are not theoretical examples! They are warnings and indicators that there has been a subtle and insidious shift from customer focus to internal focus. The costs for internal non-productive labor hours include both labor dollars and lost opportunity costs. Look candidly at the organization. Keep the sales force selling. Keep the paperwork to a minimum. Let the peddlers sell and let the administrators administrate!

Remove the shackles! Put some teeth into the notion of sales empowerment. Get the rest of the company and underemployed administrators off salespeople's backs! Too many employees think their paychecks comes from Payroll. Get out of the salespeople's way! The sales job is hard enough. Customers today clearly have the upper hand in the buyer-seller relationship. This has been so and will continue to be so. Customers have more choices than ever. Buyers now tell suppliers what they want, when they want it, how they want it, and the terms of payment. Let us work together and not make the job of selling more difficult.

Working Together!

The meeting had been called by George, the Regional Sales manager (RSM), to review the status and next steps for a large competitive sales opportunity. There were about ten people in the conference room, including myself, from various functions in the company. Finance, Operations, Training, Contracts, Marketing, and Technical Support had been asked to join the meeting to discuss if sales had maximum support to ensure this new account win. We were head-to-head against a tough competitor.

After I had updated everyone on the account's status, customer politics, and my recent actions, I presented what needed to be done by a couple of managers in the room. My Branch Sales Manager added his comments explaining the need for additional resources. Just as the RSM started to speak, one of the managers at the meeting, Bill, started speaking at length on what he believed was a better approach for me to take in order to win this business. Suddenly George, the RSM, looked at Bill and said, "Bill, have you read the sales account plan and the request for quotation we received from the customer which describes their requirements?"

"No, George. I have not had the time."

"Then shut up, Bill. You have come to the meeting unprepared and do not know what you are talking about." With that Bill shut up and the meeting went on.

Cold Calling

When was the last time you visited a company unannounced to discuss their present computing needs? Are you prepared to discuss your products and benefits on an impromptu basis? When was the last time you made this kind of foray into your territory with your manager? Does this sound silly or amateurish? Is this approach realistic for the '90s?

One day my manager at Data General, Gary, said to me, "Let's get in the car and ride around west Baltimore and look for some business." This sounded interesting. We drove around and found businesses I did not know were there. I thought we might be wasting our time. Actually we were honing our selling skills. We had some fun and were even physically escorted off the premises of one company after politely asking, "May we please speak with your data processing manager?" I wonder who sold them their last computer?

If the thought of this approach sends chills down your spine, then you are not a *peddler*. You are afraid to hear the word "No!". If I do not hear the word "no" a few times a week, then I figure I must not be working. The selling process by its very nature forces a salesperson to speak and meet with many people during the course of a sales cycle. Obviously these discussions are to get or give information. Sometimes these discussions are to seek guidance or approval for an idea or course of action. I expect to talk to strangers each week, discussing their computing needs. Otherwise, I am not selling; I am marking time.

* * *

Developing a Personal Link

When someone gives me their business card, I staple it to the middle of a 3" X 5" index card. I then write on the bottom of the index card the date we met and the primary subject of our meeting. This technique has been invaluable when I meet the same person years later. I have been able to politely remind them of our earlier meeting and dialogue which they seem to appreciate. This is one technique used to develop credibility and begin building a relationship with another business professional.

I stay in touch (SIT) with people to maintain a personal link to them. Staying in touch with people is crucial to all business people, but especially salespeople. This work habit increases my credibility with customers and builds my network of contacts and business associates. I have met people during my travels who I know will be successful no matter what they do. They seem destined to achieve success. They are organized, focused, and results oriented. I identify these people and make sure I include them in my SIT file.

The main activity I do as a sales professional which accounts for a large portion of my success is follow-up – follow-up meetings, discussions, commitments, concerns, and contacts. The importance of follow-up was taught to me many years ago by Neil Humphreys, a professor at Longwood College. I follow-up meetings and discussions very carefully. After listening carefully, I take note of what is said, requested, committed, mentioned, or desired. Then I take action or let the person periodically know I am still working the issue(s).

Follow-up is a key to success, but to maintain the long term relationship, I also use a combination of articles, memorandums, electronic mail, telephone calls, notes, white papers, brochures, manuals, birthday cards, breakfasts, lunches, dinners, and so on to make sure individuals know I am thinking of them – maintaining a personal link. I employ these various items on a one-week, two-week, three-week, four-week, eight-week, twelve-week, or sixteen-week interval, depending on the person and circumstances.

You can also read the newspapers and trade press looking for articles which mention your contacts in a positive light. Sometimes I have sent the article or a note to the person congratulating them on their success. These techniques make customers more accessible when I want to see them or get information. More importantly, it develops friendships. I often ask customers for their advice or opinions to show I value their experience. Customers of mine in the 1970s are also friends of mine today. I still see them; I still want to see them. They still ask me for advice concerning their computer-related purchases. This gives me great personal satisfaction as these relationships continue to grow. Once a customer, always a customer!

* * *

Be Punctual!

It has been said that punctuality is a personality trait. Some people have it and some do not. However, my time and a customer's time are valuable. Salespeople who are chronically late scare me as a manager. I cannot comprehend how they justify their lack of punctuality. Do they understand the message they are sending? People who are late to meetings and appointments come across as amateurs at best or simply rude at a minimum.

The subtle messages that we broadcast through our behavior often have a more powerful influence than the verbal messages we may provide in our conversations. Time spent waiting for someone to arrive to discuss an issue or to begin a meeting is usually wasted. The individual who is left waiting has been given a very clear message: "my time is considered less valuable." Don't lose the deal before the discussions have begun.

To ensure a timely arrival, I often end up arriving early. I use this extra time to scout the area, chat with other people in the organization, or do account planning. I make sure I use waiting time in a productive manner.

I worked in one company where scheduled meetings, both internal and external, would actually begin ten to fifteen minutes late as a standard; whenever the salesperson arrived was considered acceptable. This corporate culture did not value punctuality. This is inexcusable in our competitive environment today.

Do You Know the Way to Ummafoor?

A salesperson asked me to attend a meeting at the National Cancer Institute (NCI) located in Fort Detrick, Maryland. Fort Detrick is a military facility traditionally dedicated to creative biochemical research. NCI is an advanced research facility seeking advanced ways to conquer cancer. A part of the NCI is located in a building on Fort Detrick's grounds. I had never been to Fort Detrick and therefore received general directions from the salesperson. I eventually got to Fort Detrick and parked in a lot next to a double row of one-story buildings which looked like they had had the honor of being inspected by General John Pershing.

I began to walk around looking for Building 7 with no success. I became concerned because I detest being late. Finally, it was time to seek help and get accurate directions. I approached one of the elegant one-story huts and thought I saw someone standing behind a set of double doors. I did not see any signs so I walked in. I passed through two sets of double doors and suddenly found myself in a laboratory staring at a man wearing a full-dress biological warfare suit. None of the man's body, skin, or head was visible. He was wearing a sealed body suit with a head cover which had a sealed-glass face plate. The man was gingerly setting a small metal box on a counter top in the laboratory. After he carefully set the box down, he turned and stared at me. I calmly asked him how to get to Building 7 and he said, "Ummafoor, um foor, fa umm!" as he pointed outside with a heavily gloved hand. I said thanks, turned and left. On my way out of the laboratory, a few steps from the door, I noticed a two-inch cockroach lying on its back stiff as a board. When I finally got to Building 7, I kept wondering what "Ummafoor" meant and what had ruined the cockroach's day.

* * *

Be Proactive!

One of the pitfalls of getting busy is that you find yourself reacting to customer demands, management requests, and day-to-day business problems. A salesperson who is reacting to requests and events is at the mercy of someone else's agenda. They are not driving the business process. They are not in control. They are not "making things happen!" Making things happen is what sales is really all about. Stir up the status quo! Ask a tough question. Find out from your customers what your competitors are doing better. Sidestep customer internal politics, where possible. Go see the CEO or COO or CIO or CFO! Be aggressive! Do not treat policy as a god. In a constructive manner, be an agent for change!

Beware! Never accept a "no" from a person who does

not have the authority to say "yes". When someone says "no" to a suggestion, approach, or request, make sure they have the authority to say "no." They only have the authority to say "no" if they also have the authority to say "yes." Inexperienced salespeople have been sent to count hen's teeth because they took a "no" at face value. Understand the other person's position, authority, and agenda where possible. Correlate information and answers to questions where possible with other employees, always remembering to avoid internal customer politics.

Some salespeople seem to have a fear of being perceived as appearing too aggressive. Aggressive behavior to them may imply being rude or "shooting from the hip." Not true. Sales aggressiveness is doing your homework, being prepared, and being first with a new proposal or idea. This is constructive aggressiveness. Lead your customer's management through the decision-making process; do not spend your cycles guessing the direction of the "pack."

Being proactive and not reacting to events is the primary challenge of a salesperson. What does a salesperson do to be proactive? How do you keep competitors on the defensive? How do you convince the customer you are keeping their best interests in mind as you pursue mutual success? Ask yourself these questions:

- What is your competitor doing in the account? What is the basis for their success or account control? Why are they respected, even defended, by the customer?
- How can you help your customer as he or she engages in business process and system re-engineering? What can you or your resources do to help your

customer achieve a more efficient business operation? Are there new industry trends you are seeing which will affect your customer's long-term business success? Can your products and services help your customers achieve their long-term goals? If so, how? How is technological change in your industry and in your customer's business affecting their business processes?

- Do you see new business or revenue opportunities for your customers? Can you help them put a business deal together, so to speak, for their benefit?
- Do you really listen to the worker bees in your accounts who always have ideas to make your products better serve their environment? How are your products being used in other accounts which may be applied creatively to this customer's operating environment?
- Think and analyze! Can you develop your company's unique products and discriminators into a package which will move your customer to action? Do you draft a letter, submit an unsolicited proposal, or simply chat over coffee? Select an approach and take action. Drive the sales process before it drives you!

Too Much Stress!

Whenever I visit headquarters, I always take time to visit old friends, meet product marketing managers, and repay favors with a lunch here or a rose there. A *peddler* will use every opportunity to stay in touch!

I stopped by George's office to say hello. George was the vice president of sales, a friend and colleague for many years. We chatted about business, old contacts, and rumors. We always asked about each other's families since we both had difficult, albeit different, home situations. George asked me how Bob, my son, was doing. I started to give George my stock answer about Bob getting better and beginning to overcome his severe health problems, but I could not get the words out. I started crying, really crying. A lot of pain and frustration poured out. George was quiet. He did not say anything. After several minutes, I got myself together and apologized. George said it was okay; he knew it was a tough situation. We never talked about what happened. I wish I had not lost control, but I do not regret it. I guess I trusted George enough to let my real feelings show.

Sometimes you think things are under control. Stress has a way of getting your attention. Sometimes quietly, sometimes like a bucket of ice water while you are napping.

Davis' Rule Six

Competition is not an event; it is a continuous process.

The Competition

I take the competition seriously. Whenever I meet a competitor at trade shows or even at a customer site, I am always polite and friendly. I even try to come across as easygoing. This has served me well many times. Some may judge me as not being aggressive or even "missing parts." This is okay. Had I remained that unseen face working for the "other company," they may have been more aggressive.

These chance meetings have occasionally led to friendships down the road or at least contacts in a company from which I would sporadically need information. They usually seemed willing to talk about deals or accounts as long as I appeared to be friendly and casual in my manner. Obviously, this game of "let-us-trade-information-but-I-am-going-to-tell-you-less-than-you-tell-me" is a dangerous game. There were occasions when I lost the game, but at the time thought I had won. (Manage the ego!) There were other times when a chance bit of information proved to be very helpful. Still and all, I never forget — they are the competition.

I have never overtly badmouthed a competitor. This is a bad habit to acquire. Customers do not want to hear it; it turns them off. Customers expect aggressiveness and rivalry, but not rudeness or unprofessional conduct. I may attack the price or performance of a product, or a company's commitment to an application environment, but never the people or the size of the company. If a customer disparages my competitor's products or people, I feel I am being tested. I simply remain quiet or state, "I am here to discuss my company's strengths and products, and am not in a position to comment on their offerings." To my knowledge this response has been well received.

I take the competition seriously, always have, always will. I have worked for small companies for most of my career. A small company always has less resources, less documentation, limited product scope, and fewer support personnel. Nevertheless, there is no direct correlation between size and winning, and commitment to winning. Larry, a former colleague, and I beat a large, entrenched, well-known competitor in a large deal several years ago. I repeatedly beat another large company at another account in Maryland, in the 1980s. Winning does have a direct correlation to individual commitment and hard work.

Sales is where the rubber meets the road. No other company function matters if sales is not successful. I discovered that in 1975 the first time I was laid off in this industry. As I mentioned earlier, my job was to install FORTRAN-based simulation software. The Friday afternoon I received my pink slip and two weeks pay, I asked my manager, "Why?" Chuck said, "Because if sales is not selling the product, we do not need you to install it." That simple message was a powerful lesson. I decided to go into sales so I would be less job dependent on someone else's performance or the lack thereof.

As a salesperson, I continue to remember there are many families at my company dependent on what I do, say, and where I go. When you divide the total number of employees in a company by the number of the company's salespeople, you typically get a number between ten and twenty, or let us, for discussion sake, say fourteen. This means there are fourteen other families in the corporation including my own, in finance, human resources, legal, operations, manufacturing, engineering, marketing and so on who are economically dependent on my actions. They are dependent on what I do, when I get to work, when I return a call, when I decide to go home, and when I get angry about a customer's demand. I take this responsibility very seriously. Some salespeople have forgotten they are personally responsible for the economic well being of fourteen families.

When all is said and done, the other company is my competition. I do not like the competition or their products in my accounts. It makes me angry! They are trying to take food from my family. They want me to lose my home so my children have no place to sleep. I want them to stay out of my backyard. Yes, this is serious. I take it seriously!

* * *

Qualifying the Sales Opportunity: Sowing the Seeds of the Future

Careful, consistent, and deliberate qualification of opportunities is critical to the successful application of resources. Every minute spent qualifying a sales lead will save irreplaceable time and resources later on. Sales is a logical process. As with any logical process, sales requires certain steps be performed that will produce an expected result. When certain steps are eliminated or performed poorly, the probability of success is diminished.

How do you determine if you should invest scarce resources pursuing a piece of business? Returning a telephone call, mailing a brochure, or sending out a standard price quotation are all part of a hi-tech salesperson's job. These activities are routine in nature and should not impact significant business opportunities. However, when a new opportunity comes along requiring a significant investment of time and resources which can impact the outcome of other sales situations, then there must be a sanity check to make sure the business is qualified, that is, business which your company has a greater than even chance of winning.

Sales management must have their arms around the business development process and the sales forecast to appropriately prioritize the opportunities and make sound resource allocation decisions. Consistent preparation and presentation of a Sales Qualification Report, as described below, for each key opportunity provides management the information needed to make informed decisions:

1. Description of Customer and their Business Objectives
2. Description of Sales Opportunity
3. Relationship to our Business goals
4. Why/How We Will Win
5. Business Potential – short and long term
6. Competitive Issues
7. Account Decision making Process
8. Bid Approach with Estimated Investment
9. Identification of Adjunct Business partners
10. Assumptions and Risks
11. Recommendation(s)

Though most of these items are self-explanatory, some key items deserve a little more attention and discussion:

- Item 4, Why/How We Will Win: Is it realistic to expect this customer to do business with your company? What is their real criteria for selecting a new supplier? Articulate your candid insight concerning the customer's reason(s) for selecting your company over all other competitors. You will seldom know the true reason you win a deal but you always believe you know why a customer prefers your offering over all others.

- Item 5, Business Potential — short and long term: Present the initial, short-term, long-term, and related business which can be gained from this sales win. Be realistic! Expected revenue should always be more than the amount of the initial order in the world of hi-tech sales.

- Item 7, Account Decision making Process: Historically, how does this account make decisions when acquiring expensive, hi-tech products and services? Knowing the decision-making process and decision makers for key business acquisitions in an account is necessary to ensure a sales victory. This item requires an understanding of the formal and informal relationships within the account, e.g., who is on the organization chart and the influence chart in the customer's organization.

- Item 8, Bid Approach with Estimated Investment: Outline the suggested approach or strategy to bid the project and bring this opportunity to a successful close. The outline should include the win tactics: Will you bid low price? high quality? best support capability? fastest support response? The estimated investment should include a high-level list of internal and external resources including people, skills, facilities, and equipment required to appropriately support the business pursuit effort.

- Item 9, Identification of Adjunct Business partners: Will you partner with another company in a creative way to ensure you address all technology and risk areas? Identify the products, services or skills you do not have in your offering that a complementary company can offer this customer. Highlight the strengths and weakness of each potential business partner.

Once sales management has this information, the answers to critical questions about priorities, resources, competition, and the business forecast can be addressed in a proactive and holistic manner. Examples of the questions management should be considering include: What large deals may be closed in the next ninety days given proper attention now? Should we stop pursuing another opportunity on the forecast and apply our limited resources to this deal? Are these opportunities really qualified for our company? Do we have a greater than even chance of winning this business relative to our competition? Is the salesperson on this large deal driving the sales process or reacting to the situation in the account? How realistic is the sales forecast?

Qualifying new business is a Sales 101 function. However, qualifying sales opportunities seems to be a lost art these days. Many salespeople and sales managers are under tremendous pressure to simply show new account activity each month and new names on the forecast. In response to the pressure for activity and "numbers," there is a certain mind set to "chase everything." The reward structure actually punishes management for focusing efforts and resources on high potential items. This is a systemic problem.

In an environment with scarce resources, the "chase everything" philosophy can be successful only in the short term. Scarce resources will be ineffectively spread out and unproductively applied. Good, skilled professionals will step up to the challenge as best they can; however, winning is driven more by luck than skill. Given the increasingly competitive nature of the business, this short-term view will result in the eventual loss of both resources and opportunities.

Our Kind of Business!

Management had assigned me to help an account sales team pursue and win a large competitive deal. I had reservations about our ability to win this opportunity, but my manager told me I should accept this assignment. After working on this program for over a year, by assisting the sales team, my manager asked me to give a presentation on the opportunity at the next management large-business review meeting. This formal meeting provided anointed individuals positive exposure to a senior management committee. Sometimes top management would attend this half-day session. My manager said it was time for me to be "seen," besides, the deal I had been working for 65 hours a week for the past several months was worth over $30,000,000 to the company.

I prepared 15 color overheads presenting key

points on the status of this highly competitive opportunity. During my 25-minute presentation I explained:

- The background of this project
- The importance of the new system to the customer's mission
- An organizational chart displaying the account's decision makers and the decision making process
- The evaluation criteria which would be used to select the winner
- Our technical solution which included a system architecture with more than adequate expandability
- A technical and operational risk analysis of our solution compared to our competitor's alternative solution
- A description of our discriminators and how we planned to emphasize them to the customer
- A complete competitive analysis including technical and cost comparisons
- Our "win strategy" which included a plan to keep our margins unusually low while planning to substitute higher-margin technology in the second year of this deal to improve profitability. Yes, we knew our strengths and probability of success. This was our kind of business!

During my rehearsed and polished delivery my

manager was beaming confidently. A couple of minor questions had been easily answered. This was good visibility!

Just as I was concluding my remarks and was beginning to leave the front of the large conference room, a staff person who worked for a vice president asked in a matter-of-fact tone, "Bob, will we win this deal?"

"No," I said as I looked the person in the eye. The room became silent. "Why not?" queried the staffer. "The customer really does not know us," I responded. I left the silent room so the next review could begin.

We did not win this opportunity and I was never invited back to present again.

Has the sales management sanity check of ensuring new business is truly qualified for your firm been squeezed out – overlooked to a large extent? The pressure to show activity, not quality opportunities, results in poor application of resources and a reduced win ratio. Does your company measure activity or quality?

* * *

Too Much Pressure

When selling for any company you are always under pressure to sell – get revenue. This is particularly true in a start-up company environment where each order means life or death. The pressure on all salespeople to produce results, especially today, is incredible. Also, the product you are selling is new – brand new, maybe too new!

When We Meet Again...

I was in the midst of a serious negotiation with a large systems integrator in Rockville, Maryland who was looking at possibly bidding our product for a major opportunity with the federal government – a deal I felt they had a good chance of winning. I was in the Vice President (VP) of Research and Development's office with two members of his technical staff answering specific questions about our product's readiness or commerciality (to use the federal government procurement term). This means the product is truly ready to be shipped and utilized by a customer.

We had a couple of our senior managers from headquarters with us on a conference call. The integrator's VP was asking pointed questions about the readiness of the product for real application usage. Was our product ready to be used in a production environment? I could tell he was concerned and had focused on a key area of potential (bidding) risk.

It became clear to me as the discussion proceeded our people at headquarters were beginning to "sugar coat" their answers. Then finally one of our managers answered a question about system packaging falsely. I immediately and quietly shook my head, "no," to make it clear to the VP the answer he just heard was not accurate. This happened once again after another question was answered. Shortly thereafter the conference call was politely concluded. I clarified my "head shaking comment" to ensure the VP had an accurate picture of our product capabilities. The VP thanked me for my candor and the meeting was ended.

The customer did not bid our system. We offered too much risk; our product was not ready for day-to-day usage. They won the deal later using Stratus Computers. I do not know what the VP thought of my company or me. We never discussed it. Clearly my goal with him was to maintain my integrity and credibility. I knew I would be calling on him again one day even if we both changed employers.

A product can be too new when it is not ready for day-to-day commercial usage in a real business environment. Customers often want leading-edge not bleeding-edge technology. Customers do not want to be guinea pigs for an overzealous salesperson or company. Occasionally a customer may become an unwitting, unofficial beta site for product that is not really ready for production workloads. It is unlikely this same customer will be willing to discuss business with you in the future.

Evaluation Criteria

After two and a half years of work, I finally won a $120,000 order from a new account. During this protracted procurement for a relatively small amount of money, many of the original competitors had either dropped out or been technically eliminated. The deal came down to two companies: mine and one other company. We both had similar products, prices, and support. Either company could do the required tasks. Why did my company win the deal? At the time I believed it was because we had offered better price and performance.

At the time I wondered, should I have continued to spend so much time pursuing such a small order? Yes, I had other business coming in and did consider this account to be strategic in nature, because the customer's application was unique and state-of-the-art in the field of maritime simulation. Was I too persistent in pursuing this business? What other opportunities did I miss?

Several years later I got the chance to ask the key decision maker, at his retirement dinner, why he really chose my system over my competitor's. Mack said, "Because you kept coming back and I believed you were more interested in what we were doing." After all those years, I discovered the real reason I had won this deal. I believe we seldom know the true reason(s) we win a deal. There are always unwritten evaluation criteria.

The Better Mouse Trap

I was concluding a 45 minute presentation to eight people who worked for a large, well-known company. I had presented the capabilities and benefits of my company's computer system. I explained how this system would be an ideal platform to off-load a specific set of applications from one of their mainframes. All of the customers had technical backgrounds and lots of experience delivering computing resources to end users.

After a couple of questions related to response times, I was asked to configure a computer system appropriate for their application environment. Once this was finished I asked if there were any other questions.

"How much does the system cost?" asked one of the attendees.

"Including system software, installation, and one year of maintenance, about $90,000," I replied confidently. I knew my price for the full system plus maintenance was significantly less than the current maintenance-only costs on the existing systems.

There was a brief pause, then two of the managers in attendance started laughing. As they stood to leave, one said, "You can't sell a system of that size for that price!" With that, they left the conference room. The other attendees said nothing and remained seated. I was asked a couple of polite questions then the meeting quietly ended.

I did not get the sale. I do not believe the real objection was the price. Looking back, I believe they perceived risk. The risk was greater than any cost savings that could be gained with the recommended system. There was too much risk for them to feel comfortable with a smaller supplier. I had failed to anticipate this concern and address it head on.

* * *

Sales Channels

In this era of micromargins, if you are selling a box with a unit price less than $50,000, the old way of many sales calls, management meetings, and lots of customer hand holding is no longer profitable. Manufacturers can no longer afford expensive and extensive end-user sales forces. As a result, many computer companies today have moved to indirect sales channels instead of a direct sales force. The companies are hoping this change will allow them to eventually move fifty percent or more of their product volume more efficiently. This is a risky presumption.

The simplified view of commercial information technology marketing channels displayed in *Diagram A* clearly demonstrates some key distinctions for sales in the indirect selling model:

- There are now multiple paths between the manufacturer and the ultimate customer.
- Selling through an indirect channel, be it VADs, VARS, OEMs, SIs, resellers, and so on, introduces at least two sets of key decision makers, buyers, approvers, recommendors, business issues and agendas into every major sale, not forgetting consultants.

Commercial Information Technology Channels

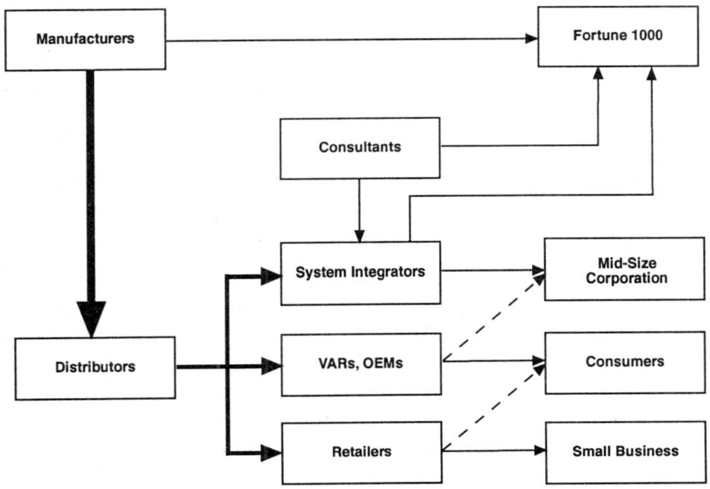

Diagram A

- Consultants are at the top of the information technology food chain – they sell expert knowledge and relationships. Their influence and importance is expanding. They command significantly higher margins in the services business than do manufacturers. They will survive the continuing march of the process of commoditization (which I will address shortly).
- Competition is intense at all levels and channel relationships are not sacrosanct.

The move to indirect sales channels has introduced a variety of issues and complexities not originally envisioned by the manufacturers. Account management, building customer loyalty, account control, and product support are all more challenging when products are sold through indirect channels.

Some salespeople like commercial sales versus government sales, end user versus system integration sales, or end user versus distributor sales and so on. The bottom line is, each market segment requires unique selling skills and abilities. Selling to end users versus selling to resellers who, in turn, sell to end users is a different selling model with different attributes than direct end user sales. Selling to resellers who are in turn selling through a variety of contractual vehicles to the federal government adds another dimension to the complexities of indirect account management.

However, many companies in the transition to alternative sales channels have simply assigned end-user salespeople to "manage" one or more resellers on behalf of the company. These salespeople are now measured on a combination of direct and indirect sales revenue goals. This mixing and matching channels with a sales force trained to sell to end users has had mixed results to say the least.

Selling is Selling – Right?

Genny, a salesperson with my company, poked her head into my office, took an exasperated look at me, then walked in. She dropped into the chair by my desk and gave me a weary look. I asked if she was okay. She said, “I am tired and frustrated.”

"I have been calling on system integrators for two years and I do not understand what they do. End user sales I know, but these guys ask for the weirdest stuff!”

“Have you discussed this with your manager?”

“He’s no help. He doesn’t really understand what they do anymore than me.”

“Have you talked to anyone in Training?”

Genny looked at me and started laughing. I told her I was willing to help. I suggested we meet in two weeks and discuss the differences in selling to this market segment versus selling to end users. I asked her to call me to schedule the time. She said she would call me. I never saw her again. She left the company two months later.

Getting and keeping the mindshare of a distributor or system reseller who also sells ten other (overlapping) product lines is a skill unto itself. This indirect sales channel has seen hardware companies come and go, their suppliers' salespeople come and go, and of course promises of new products which ship later and later, if at all. Why should they get excited about another marketing pitch from your new VP of indirect channels? Keeping the reseller's management, sales and telemarketing force, and technical support people enthused about your box is most challenging because they can always sell the other guy's product if they see a way to a faster sale. Consistently selling products through indirect channels is a harsh business and not for the faint of heart.

Also, this indirect sales model also has some inherent risk areas for the manufacturer. You may be able to sell modems and PC add-on boards without too much support infrastructure, but system sales, be they large-scale servers or WAN networks, require continuous support, direct and indirect, from the manufacturer to the end user of the product. You must distinguish between "board" sales and system sales. System sales require seamless support, consultation, and hand holding regardless of who actually receives the purchase order. Are your channels or resellers providing the necessary level of support and commitment for your system level products? Since they rarely have the same level of passion for your products as you do, how successful will they be? It may be time to re-evaluate your company's channels strategy.

* * *

Winning the Business is Just the Beginning

Often when a major new account is won, sales management hears about the potential for millions of dollars in revenue expected from the account during the next few years. Their eyes glaze over. They raise the sales forecast, order their new BMW, and hire another salesperson to "help" you with this new account. But for some reason the revenue never rises to management's expectations. "Why?" they lament. Is it possible that they have taken the business for granted since the victory party and assumed the revenue would now just flow?

Once the business is won, it is not time to step back and start counting the profit. Now is the time to observe and carefully assess the trends in the marketplace. It is the time to start building customer loyalty and strengthen executive level relationships. It is time to analyze, track, and monitor the competition. Existing accounts need as much TLC as new accounts; maybe more, since they are now part of your installed base, already in your camp.

I remember having accounts in the past which were mine. My company had, to use the old term, a "lock" on the accounts. Competitors were wasting their time trying to penetrate my key accounts. These days are long past to say the least.

There are forces at work which cause competition to be a never-ending process in today's dynamic marketplace. These trends can make your sales victory less secure with each passing month:

- The continuous advance of technology and technological standards. The products you sold the account may not be as effective or needed next year simply due to other technology, e.g., PC-fax to PC-fax instead of facsimile machines sending to other facsimile machines.

- Business process and information re-engineering, under the guise of many names, will go on and on. As a result, a company function or department who is today your best customer may not exist next year.
- Key decision makers in the account loyal to your company may leave or change jobs. There is continual turnover in management positions in this industry, as much as 50% every twenty-four months according to one source. Chances are low you will continue to call on the same decision makers twenty-four months after the initial sales victory.
- Your customer's competition will force them to take action to prevent loss of revenue or market share. Companies implementing new technology into their redefined business processes will force changes, sometimes radical, onto their competitors. These changes may in turn cause large increases or decreases of investment in a given technology by a specific account.

Given this volatility, today you are vulnerable to attack before you receive the first or second purchase order from a new account. The days of having a "lock" on an account are over. New accounts are not fiefs. They can be taken away by an aggressive competitor. Or they can be lost through inattention.... Who is watching the store?

In developing our sales account plans, we are constantly told to promote and encourage consistent executive relationships. You must have multiple relationships into the account reaching different levels and functions of the organization. Multiple relationships and communication paths into an account will provide you with a means to correlate and verify data you receive concerning a given competitive situation. This feedback is crucial for keeping your competitive weapons focused on winning new business. The term account relationship itself implies management-to-management dialogue at multiple levels is occurring. If my company wants a relationship or partnership with a potential account, I and my manager must know more than one person in the account, that is, have more than a single point of contact. This way long-term business relationships may continue in the account even when people in either organization change.

Repeatedly during my sales career I have been asked or told to schedule a breakfast or lunch meeting to foster a relationship with a key executive in my key account. Later, more often than not my management was unable to attend these meetings due to scheduled internal meetings. Building customer relationships is a continuous investment. It is unrealistic to expect all of the revenue to be there when executive management relationships are not in place and the above four trends are not anticipated. This inattention can provide the competition an open door.

Can you surmise your competition's goals this year for your key accounts? Against whom do you usually compete? Who are your toughest competitors? How or why have they won a given deal? Do they have assets or skills you do not have? What are the strengths and weaknesses of each primary competitor? Who are potential competitors? Who sells a similar product? You may rate your competitors by analyzing these variables:

Market factors

- Product quality
- Company image
- Product differentiation compared to yours
- Breadth of their product line in terms of customer needs and wants
- Customer and application foci
- Usage of distribution channels
- Types of advertising, promotion, and choice of media
- Service organization
- Aggressiveness of sales force

Management factors

- Management experience and integrity
- Knowledge of business process for a customer's application environment
- Flexibility
- Risk takers
- Strategic and tactical planning skills

As Steven Milunovich of Morgan Stanley said, "Selling is about competition. Selling does not occur in a vacuum, but rather in a complex environment in which rivals fight to steal customers and every action results in a reaction, much like a game of chess." Be prepared.

Is This the Time?

To ensure we spent time together doing something my younger son wanted to do, I became an assistant coach on his baseball team. At the very first practice, I recognized another boy's father, but could not quite recall why I knew him. After a couple of weeks, I determined he was the deputy program manager on a large, complex procurement for a federal government agency. I was interested in this procurement and wanted to ask him some general questions.

However, time with our children is special. I was not comfortable discussing business with him before, during, or after the games. I was not being coy. I felt I was respecting his time with his family. Although I badly wanted to hear his opinion, this was not the place and time. Furthermore, it may not have even been proper, since he was a procurement official of the federal government and cannot freely discuss business with one vendor to the detriment of another. So, I kept my mouth shut and did not cross a line I had drawn in my mind.

Davis' Rule Seven

If there is no sense of urgency, then the organization is drifting.

Sense of Urgency

Have you ever walked around the office at 8:00 A.M. to sense the pulse or atmosphere of the organization? How many employees are in the office? How many are working or simply chatting over coffee? How many are planning their day or reading the paper? How many employees take 45 minutes or 75 minutes for lunch or eat at their desks? Do meetings start on time with written agendas? If the boss is out, is half the office out too? Do unannounced visits by the Vice President cause nervousness? Are there lots of sick days taken on Mondays and Fridays?

The answers to these questions may reflect a lack of leadership, low morale, poor internal communication, excessive workload, or of greater concern — no sense of urgency.

An important differentiator between winners and losers in business today is a sense of urgency. In a winning environment:

- All employees are focused on the importance of each customer, each order, each telephone call.
- Customer deadlines are of greater importance than internal schedules.
- Employees are motivated and rewarded for new ideas, winning new business, saving money, and promoting the image and success of the company.
- People are returning telephone calls to both customers and other employees the same day.
- Sales management is from successful sales backgrounds.

- The company's customer satisfaction programs have real teeth, genuine rewards, and appropriate resources.
- The sales forecasts are being submitted on time with real numbers.
- The forecast system itself is credible and seen as adding value to the selling process.
- The forecast can be used to plan inventories and workload with a high degree of confidence.
- Expense report costs are being kept to a minimum.

How does your organization stack up against this list? Be honest. If you cannot clearly equate these actions with your organization, the company may be drifting.

Management's greatest challenge is to maintain an environment to motivate all employees. Management must instill a sense of urgency in their organization and provide leadership to ensure all employees support this notion. It is like being a professional sports coach with a season that never ends. It is not a matter of berating and threatening; it is a matter of leadership, and never forgetting the importance of the customer. All company functions and groups must be striving for the "first string." Sales cannot be charging down the field with the ball with slow blockers, no waterboy, and coaches still in the locker room.

If there is no sense of urgency, then the organization is drifting. When an organization is drifting, it is losing market share. It is letting its competitors take the lead in its customer's mindshare. Forces of the marketplace grind steadily and will overcome corporate cultures where complacency and burnout are rife. A company which is drifting is waiting for change to happen – negative change. Reorganizations, layoffs, benefit reductions, demotions and so on will happen to all levels of a drifting organization. Beware, is your company drifting?

Lifting a company out of a downward spiral, while revenues are down and layoffs are ongoing, requires three elements. First, management must have a vision and strategy in their minds which has to be clearly presented to all employees. Each employee must be able to see the path to success as espoused by senior management. Are the employees believers? Too often I hear top management say things for the benefit of Wall Street, all the while wondering if customers accept what is being said, let alone employees.

Second, the company must have the resources and talent to execute the strategy envisioned by the leaders of the organization. Waiting until the stockholders are screaming with their thumbs pointed down and little money is left in the till is too late. Why does top management wait until the walls are crumbling down around them before they have sudden revelation it is now time for change? Change will occur with or without your cooperation. New ideas must be considered when all seems well in an organization; change must be anticipated. This is the primary responsibility of top management. Remember, if you are not moving forward, a competitor is preparing to pass you.

Last, there must be a commitment from all of management to support the organization's metamorphosis. A few executives at the top cannot do it alone. There must be buy-in throughout the organization. The challenge of turning poop into pearls requires everyone at their duty station giving an extra measure of effort every day. The extent of the complacency of middle management in most large organizations I have witnessed is incomprehensible. They did not get where they are by rocking the boat. They seem secure in their jobs, rarely taking risks, never saying no to their boss, never demanding a customer problem be addressed beyond the minimum response required, and so on. These innumerable drones are the ones who must follow through and personally make sure the new strategies of top management come to fruition.

Scary, isn't it?

Don't Sue God

I received a call from our Accounts Receivable (AR) Department who complained about an account in my sales branch terribly late in paying their bills. AR had sent strongly worded letters and made telephone calls to the account demanding payment and received no reply. It was now time to take stronger action.

I asked AR for more information about the account since they were not familiar to me and quickly discovered this was the Regional Headquarter offices of a well-known church. AR said, "We must go after them aggressively."

I said, "No."

"Why not?"

"I am not in the business of suing God."

> There was a pause on the other end of the line. I suggested we give the church more time, which we did. They eventually paid all their bills. Can you imagine the publicity had we taken stronger action? God forbid the thought!

Customer Satisfaction

Large companies often make it hard to conduct business with them. They have evolved complex systems and procedures that treat customers virtually the same whether they are buying a pre-packaged $100 manual or a $35,000 system. The complex processes that have evolved to purchase simple products exemplify the internal focus of a company – not the customer focus needed to sell "commodity" products in today's business world.

If I want to buy milk, I want to go to the store, get the milk quickly, pay a fair price, and get out of the store. I want the entire process to be efficient and quick. I expect little interaction with store employees and when it happens, for them to be polite and knowledgeable. Yes, milk is a commodity.

Today, when buying a commodity product, like a memory board or a modem, the process should not be much different. A customer expects quality and efficiency – no more, no less. The customer may pay more for higher quality. The key is that the transaction is quick, and interaction with all employees is polite and productive. This is all common sense. But too many companies have gotten away from these basics. The basics: keep a business transaction simple, efficient, and customer focused.

Last year I stopped by a fast-food restaurant for breakfast in Washington, D.C. On the wall near the cashier in plain sight was a fascinating poster. This poster presented all of the opportunities available to an employee to impress a customer who had come to the restaurant, for example:

- Entering the restaurant – Is the facility clean? Does the environment seem "warm," feel friendly?
- Placing an order – Are the employees courteous? Do they listen carefully?
- While waiting for food or in line customers always observe employees – Are employees efficient in performing their duties? Do they show they care about the customer's order?
- Getting the product – Is the food handled properly? Is the order filled correctly?
- Using the product – Does the product taste good? Does it meet the expectations of the customer?

Customer satisfaction is the focus of this poster. It's purpose and objectives are clear to all employees. Management obviously wants their employees to be prepared, to understand there are many opportunities to ensure customer satisfaction from a single sales transaction.

Do your employees understand all of the opportunities during a sales transaction to ensure customer satisfaction? From the first telephone call to your company, to the day the product is deinstalled, where can your employees excel in serving the customer? Are they sensitive to all the times they are being observed by the customer? Are your employees focused on the customer or on internal processes? Do they really believe the hackneyed phrase, "The customer is Number One!"? Does management?

* * *

Work Ethic

Strong work ethic is a topic you seldom hear about today. It seems this concept applies to the older generation, or people over 50, ha-ha. Anyone working in any job today should be committed to winning – accepting the responsibility to do the job well until success is achieved. This means taking ownership of a task and ensuring it is completed, not only properly, but beyond what was asked or was expected, when possible. Too many people today seem to be going through the motions, doing just enough to get paid every two weeks and never upset the boss. Go with the flow. Do not take a chance. Ignore something you dislike and it will go away. In this age of downsizing and layoffs, this work philosophy by some employees has cost other dedicated, hard-working employees their jobs. When the invariable layoffs come, the quiet, go-with-the-flow employees are not the first to go. What an interesting paradox.

Many employees today comply with management's direction and suggestions. They obey the boss and make sure they follow company policy, no more, no less. Organizations today do not need compliance; they need commitment. Commitment means I can add value to my manager's vision and will do what it takes to make that happen. If all employees are not committed to doing the job right, helping their company's salespeople win again and again, then the organization is drifting. This lack of pride and commitment to getting the job done will be obvious. When the inevitable layoffs begin, these are the people who will often survive – the people who know how to sell better than the salespeople, but who will never consider carrying a sales quota. They spend their work day managing their careers and saying yes to the boss instead of taking ownership of a complex, unpleasant customer problem. You cannot sit on the fence forever. As I have said before, continuing a low profile in an organization will one day become your career albatross. Your department will eventually be reorganized under a new manager who may be results oriented and find your low profile nature to be unacceptable in the new "lean and mean" organization.

Our employers pay us for an honest day's work. This is normally more than eight hours in sales. Hi-tech sales is not an 8-to-5 job. Anyone who thinks or treats sales as a 8-to-5 job is cheating themselves and their company. Responsibility for customer accounts is not forty-hours-per-week duty. No matter where I am, I call in for messages and stay in touch at least three times a day. I have called the office from vacations, my honeymoon, and remote state parks. Salespeople who go on a two-week vacation and never call the office amaze me. They must have total account control and do not have to worry about the actions of a competitor. I tip my hat to them and hope I can compete with them soon. Any company who has an environment where forty hours a week gets the sales job done, or so it seems, is fooling themselves. Listen up! There are well-paid salespeople today not giving their company an honest week's effort in spite of appearances. Does your well-paid sales management know who are the walkers and who are the talkers among their subordinates?

* * *

Take Control of Your Career

While I have worked for several companies during my more than twenty-one years in the industry, a couple of times I changed jobs on my volition, and three times this decision was made for me. All in all, I did okay financially and was able to move my career forward, following a career path. I learned to pay attention to business activities and events happening around me. I learned to consciously evaluate my career progression once a year and take an in-depth look every four or five years. After five years in the same company or job, you had better be moving up or out. It is too easy to become stagnant. If you are not moving up or out, you are coasting. If you are coasting, your career is not advancing and your future economic value is waning. No one tells you this! It is too easy to get caught in the slow seduction of the same job in the same company in the same location. A false sense of security evolves over time, causing you to let down your professional development guard.

This fear of career stagnation has forced me to change jobs when it appeared to others to be an illogical decision on my part, and not well thought out. I also know now when you leave a firm after four, five, or six years, leave while you are on top. When you have the choice, never leave while you are under the gun, failing to produce, on management's bad person list, after a fight with the big boss and so on. Leave when you are successful. Leave when your resignation will surprise most people. This way you will not be viewed as a quitter. If rumors abound, they will be complimentary, not detrimental. This will keep the door open to past associates you will meet later on your career path. They will remember you positively and as being successful, not as a quitter who walked away when the heat was turned up. Even in this age of indiscriminate layoffs and heavy competition for each career position, try to do this when you are able to.

Someone reviewing my resume would wonder why I seem to have difficulty holding a job. True, I was laid off three times and I voluntarily changed jobs a few times. But overall my career progression is the result of a plan I developed many years ago. Again, if you have had the same job for four or five years, you need to take a hard look at where you are. If you are not moving up and continuing to be challenged, you need to consider moving on. I have made career decisions with this objective in mind. I have five years direct hi-tech sales experience, five years hi-tech sales management experience, ten years business development experience, and am beginning my second year as a program manager. While this experience has been gained while working for several companies, I have not lost sight of ensuring my professional growth.

The traditional model of after high school, going to college and then getting a good job is rare and almost non-existent these days. People are living longer and working longer. You still might retire after twenty or twenty-five years with many companies. After this, are you going to start a second career? start a business? change industries? go back to college? or live out your days on your pension? Laws and programs exist or are now in development which support an employee working for a company five or ten years and then changing jobs. Retirement benefits are now vested after five years. Fortunately, new health care legislation will provide a safety net of health benefits for those people between jobs caught in the pre-existing medical exclusion nightmare.

Today, young professionals are working and learning two disciplines well. For some it is economic need; others are preparing for early retirements; and some just want the challenge. Whatever the reasons, people are going back to college or trade schools, or working two jobs. Could a person in the near future be working two professional jobs forty and twenty hours a week or thirty-five and thirty-five hours a week with benefits? There is a massive retraining need today for people 50 to 55 years old. There will also continue to be employment dislocation of white-collar and blue-collar workers in the future as companies right size and re-engineer to compete in global markets. This will force an emphasis on:

- Continuous education – Constant technical and business skills training will be the norm. Employees receiving less than four person-weeks of training a year will be taking risks – risks with their career and risks with their position in the company.
- Two subject matter expertise – You will begin to see people who are trained and experienced in two disciplines. Fields treated as separate careers are

now merging, like engineering and biology or database and network security.

- Individual responsibility for career development – Take charge of your own career development. Corporations will still offer some training and make sure a bullet called career development is placed on the agenda for the monthly human resources (HR) meeting. But by and large your career development and progression is your challenge. Ideally the responsibility for professional development is the company's and the employee's, each having fifty percent. However, as employees get buried on long-term projects, and management ramps up for the end of the quarter Big Push, this subject is relegated to the monthly HR meeting, if at all.

Manage your own career. Recognize that working for multiple companies through a career maze may be the new norm. Understand that changing jobs voluntarily takes fortitude and courage. It is difficult to say goodbye to friends and dive into a new environment. You are not just getting a new job; you are getting new products, new management, new applications, new measurements of success, and sometimes new jargon. This is good. It will increase your value as an employee. It will increase your knowledge base. Career growth today requires saying goodbye.

* * *

A Career Move: Large Companies

I realized in the late 1980s the computer industry was heading for trouble. There were too many suppliers and deep price cuts had become the rule. I also noticed some senior salespeople not making their quarterly quotas which told me the business was simply not there as it had been in earlier years.

During this time, I had a marketing job where my duties were ill-defined and I seldom had substantive conversations with my boss. When I asked questions about the future of the organization, I got polite answers, so to say, "Do not worry about things." When I asked about the need to establish relationships between our senior managers and customers, I got polite answers, so to say, "Do not worry about things." So, I decided it was time for me to worry about things and get into a new environment which might offer some direction until the chaos of the industry settled down, if ever. Therefore, I decided it was time to go to work for a big company.

After working for small companies, joining a company with over 120,000 employees was a culture shock. My oh my....I was now a small fish in a big pond. This was a new world to me. The learning and assimilation process began immediately.

I initially discovered the forms, policies, and procedures that are staples of large companies. This infrastructure of paper and policy is useful to understand and learn from a career development standpoint. This bureaucratic infrastructure has evolved over time and represents the combined knowledge and experiences of many professionals. The shear volume of corporate information and options available added a broader dimension to my knowledge base. Also, working at a large company afforded me more opportunity to work in a variety of application environments, and on large business deals. These are good, positive learning experiences.

However, there can be a component of the assimilation process which is dangerous to your career progression and the long-term success of the company – seduction into complacency. In a large company it is easy to go with the flow. You learn early: do not rock the boat; do not ask management tough questions; do not grab authority to get the job done; always check with the boss before you make a decision. There are times when this path is logical and prudent. There are other times when a salesperson needs to rock the boat and irritate management of non-sales functions in the company to win competitive business. A salesperson needs to recognize when the low-key approach is the smart choice or simply the easy choice. When a low profile becomes a work habit for a salesperson, beware....

It is easy to maintain a low profile in a large company. Keeping the reports and memorandums flowing, returning certain telephone calls, attending meetings, saying yes or nothing in meetings, and always smiling when you see the boss are time-tested ways to appear as a loyal team player. However, the team does not always win. Who is accountable when vital business is lost? Most companies would hesitate to fire the entire sales team after a painful loss. There is safety in numbers, so it would appear — but safety in numbers is only an illusion. We are all responsible for managing our own careers and W-2s.

I realized one day while working for this large company that I was slowly sliding into the 20-year seduction.

My benefits were okay. There was not too much pressure on the job. I wrote my own performance appraisals. I got a paycheck every two weeks. I also received an annual statement telling me how big my pension would be when I retired. It all seemed so cozy and yet did not feel right. Things did not feel genuine. There was a false sense of security.

There were other signs that something was tainted in paradise. My company had been trimming headcount quietly and not so quietly for years. There was too much supply and too little demand for our products for the foreseeable future. Health care benefits were eroding steadily each year. Other employers seemed disinterested in interviewing people who worked at my company.

The combined result of these signs caused me to objectively re-evaluate the real stability of my own job as well as the company's viability in a few years, particularly by 2020 when I planned to retire. I could ill afford to have the "Angel of Layoffs" pay me a surprise visit one day, scythe in hand. As a result of this evaluation, I decided I needed a job change....

When there are days with no competitive heat, something is wrong. Pressure and competition in business are ever present. When you have not stood out, when you have not fought a good fight to win more than your share of the business, when you have not dragged management into hand-to-hand combat to win a competitive deal, your low profile can eventually become a career albatross. Are you being seduced into career complacency? Have you kept your professional development on track?

Mission: Impossible

I decided to stop by Roger's office. Roger was a support manager for our sales organization. He had a way of saying things which at first did not sound important, but later on seemed profound. You had to be careful though; Roger could get you into an intelligent conversation about designing microprocessors, pulsars, and C++ in the same conversation.

This particular day, after a long pause, while I was standing in his doorway, Roger said, "Do you know the company this year has to sell all it sold last year, plus some, every year, forever to stay in business? That cannot work." I laughed and said, "Of course, Roger. Everybody knows that!" Well, do they?

If you have to sell your entire installed base plus fifteen percent, over again, each and every year you are in business, are you prepared to do it? Are your new products coming fast enough? Are you continually training your people? Are you reinvesting into facilities and new business processes fast enough? Are you beating your competitors often enough? Are you growing your market share? Is your company allocating its profits and productivity gains effectively and according to whose agenda? Are you really prepared for the future?

Davis' Rule Eight

The longer a company has been in business, the more the company's culture will oppose the changes necessary for competing for business.

Resistance to Change

There is an old Chinese proverb which says if you want to curse an enemy, give him twenty years of success. I believe many companies today are now experiencing this "curse" after all of their success in the 1980s. Today, change is a constant in the business world. Nonetheless, these companies are stuck in the old "proven" models of yesterday. What brought the company success in the past is not what will bring success in the future.

Management at all levels seems to recognize this issue. There is an attempt to introduce new ideas and prompt alternative action when they say "We must change the way we do business" or "We must develop new approaches." However, too often the spoken words are merely clichés. The words need to translate into action for true effect. It is almost impossible to inject new ideas into an organization, particularly a large company. The energy expended to implement these new ideas rapidly becomes dissipated within the hierarchy.

When I went to work for my first large computer company, I first thought my ideas could have been described as trying to single-handedly alter the course of a ship fourteen hundred feet long moving at a speed of twenty knots. I discovered the process was more like trying to get one thousand canoes going in the same direction, at the same time, and at the same speed.

When I have decided to carry the banner of change into battle with middle or upper management, I often saw a lack of understanding of the situation from my viewpoint. I do not know all of the variables management must consider, but they seem to have trouble listening objectively to suggestions, particularly to an approach used or avoided by another company. There seems to be an underlying fear of appearing not in control. Our discussion is usually brief and one sided. I get a nod of the head or a polite comment. Sometimes I feel like a Buddhist in the Vatican. Why does management seem to look everywhere for new ideas except their own subordinates? If their boss gave them the same suggestion, they would hastily agree and take action. This is not a healthy environment for developing dialogue to foment new ideas and positive change.

Implementing change in a environment of one thousand canoes requires proactive leadership and vision at all levels. Unifying all parties is the most difficult step. There is universal opposition to change. The general philosophy is "If it's not broken, then don't fix it!" But this is the basic problem! One person thinks it is working okay and another thinks the business is close to "crash and burn." Many employees keep believing all is well simply because the paycheck always comes every two weeks. When a company is in an ongoing mode of reorganization and layoffs, I personally think the environment is closer to crash and burn.

As a company grows, a culture develops that includes a certain amount of corporate memory and history of previous success. This background can create a "business-as-usual" creature who lurks the halls and wallows in corporate politics. Over time some managers are transformed into mental couch potatoes. Company politics are allowed to evolve to a fine art. These are indicators of a lack of leadership from higher management.

There is an inverse relationship between the level of politics in a company and maintaining a lean, competitive organization. Politics are normal wherever there are people. But when they supersede business matters, the company has lost control of its ability to drive the business process and therefore manage its direction. It finds itself in react mode – reacting to customers, competitors, government regulation, and an ever-changing marketplace.

Building momentum to stay lean and aggressive, to fight for *every* customer's business is the job of salespeople, sales management, and the entire organization. When non-sales functions do not share this vision and fire, then the sales job becomes impossible. When the sales organization is mentally collecting checks, then the "Angel of Layoffs" is being sent a written invitation to visit.

But who wants to stand up in a meeting and tell top management we are headed for crash and burn, let alone put it into writing in a report? Everyone wants to be viewed as a team player. An individual wants to be viewed as one who has a positive attitude and understands the company's problems. Meanwhile, the layoffs continue.

And as the layoffs continue, the stream of valuable, skilled employees quietly moves on to other companies. How many successful, hard-working employees have to leave organizations before upper management realizes and acknowledges they are really ready to listen — earnestly and openly? Sometimes top management has to bite the bullet and clean house, which they usually avoid doing until forced to by an irate board of directors or angry stockholders. By the time most executives are ready to implement the new order, their most talented people have been long gone. This helps to continue the downward spiral. Yes, we are going to see more companies crash and burn or be absorbed in what are termed as "friendly mergers."

Organizational Styles: Peddlers vs. Administrators

Usually in the early history of a successful corporation, the members of the sales organization have a reputation for being aggressive, innovative, entrepreneurial, and maybe even overzealous – real peddlers! Over time, the organization evolves and implements policies and procedures to control and manage the wild ride created by explosive growth. A new style is invariably introduced into the sales organization – sales administrators. The *peddler* and the *administrator* represent organizational styles, not specific positions within the organization.

Usually in a mature company, the organization has evolved from the "wild days." Though the good-old-days are viewed fondly, high value is now placed on getting along and developing into a smooth operation. That professionalism too often translates into placing administrators in control of the operation to manage the chaos. Sales administrators are dangerous people when incorrectly placed in an organization. Sales administration functions, which focus on the internal processes required to capture business efficiently, are important to the long term success of the company. However, the front line sales positions, by definition, require a strong external customer focus with a willingness to aggressively compete for business. The *administrator* focuses on the internal business environment. When sales administrators infiltrate front line positions in sales, this is a real problem.

Exhibit F highlights the major differences between sales *administrator* and *peddler* work attributes. Administrators and peddlers will not normally work together as a team. They are different trees from different forests.

Peddlers do not usually seek jobs as sales managers. They have low tolerance for politics. Administrators hope for management jobs so their difficulty adding value to the selling process is less visible to upper management. Administrators by their nature are not comfortable in face-to face meetings with sophisticated customers. Peddlers thrive on the challenge of constant customer interaction.

Work Attributes	
Sales Administrator	*Peddler*
Lots of meetings and conference calls	Avoids meetings and conference calls
Inwardly focused on company activities	Outwardly focused on customer activities
Company management takes priority over all other issues	Company management is merely a resource to help close a deal
Produces reports and numbers concerning the big picture	Produces reports pertaining to closing business
Focuses on budgets and expenses	Focuses on revenue
Focuses on politics	Focuses on results
Ensures time is spent with own management	Ensures time is spent with customer management
Aware of the business process and expects to be part of a win	Knows the business process and what has to be done to win

Exhibit F

The challenge for the company today is to understand who is really staffing those front-line sales positions: peddlers or administrators? Peddlers have the following personality traits:

- Communicator – communicates clearly; listens carefully.
- Competitive – determined to win; persistent.
- Entrepreneurial – can envision new opportunities; takes risks.
- Results-oriented – gets consistent results; achieves goals.
- Aggressive – assertive, not obnoxious.
- Creative – presents new ideas and approaches.
- Healthy ego – confident, not arrogant.
- Ethical – maintains honesty and integrity in all dealings.
- Leader – sets an example; gets others involved in achieving success.
- Proactive – can make business happen!
- Tenacious – persistent; does not give up easily.

Does your interviewing and screening process identify these attributes in your new salesperson hires? The first four traits are relatively easy to identify during the interview process. The sales candidate should be able to show evidence and examples of these four items. The remaining traits are more intangible in nature and may be observed over time. Are these traits encouraged and rewarded by the company?

Sales administrators try to act like sales managers, but in fact may actually reduce the probability of success of the sales team. Sales administrators are adept at managing their careers to the detriment of adding value to the selling process. They quickly learn how to package data so it is well received by their upper management instead of presenting quality information to management that may more realistically describe the competitive situation. Administrators do not necessarily present what has to be done to win the business. They present data in a form acceptable to their management, mindful of company politics and organizational issues. When the sales management positions are also staffed with administrators, this lack of quality information is not noticed. Simply put, sales administrators get in the way of closing business. However, if the company has created an internally-focused infrastructure of management and administration that rewards politically correct behavior, the final tactical decision usually has nothing to do with reality anyway.

I wish senior management, who often sees many smiling management faces at sales meetings, would line up the sales managers without prior notice and ask them to answer a couple of questions:

1. Name the customers they have met within the past fifteen days, and

2. State the purpose of these meetings and what objectives were achieved at the meeting(s).

I think some of the weenies are in for a surprise. Some of the sales managers have not met with or given a presentation to a customer in so long, they are hesitant to do so. Do you think I am exaggerating? When will senior management begin to weed out administrators from their sales management ranks? Why do they keep waiting for ninety-pound weaklings to make the big plays? Does company management have the requisite knowledge and experience

WAKE UP! Let the peddlers peddle and let the administrators administer! And most importantly, understand where their skills are appropriately applied. In today's competitive environment, no company can afford the luxury of creating an alternate reality. The marketplace is quick to realize when the emperor has no clothes!

* * *

Rewards

Many companies wrestle with the age old problem of how to fairly recognize their successful salespeople. Some salespeople have come to believe returning a telephone call, mailing out a brochure, or meeting someone for lunch is selling. This may or may not be selling. Real contribution is effort which advances your company in the decision approval process and leads to successfully closing the order. Management can no longer afford to reward mediocrity or basic activity. Reward must be based on real contribution and effort by specific individuals.

Our world is a world of ever-changing goals, priorities, and competitions. Like pro-ball, step up to the line and be ready for each play or get out of the way. We are paid to do a job and get results when the money is on the line. The boom times and cherry-picking days of the past are long gone. There are fewer players being selected for the team roster today. Measures of success were traditionally based on quantitative factors such as producing consistent numbers, or winning new business accounts. Surviving mergers and layoffs, and meeting one's financial objectives are part of the definition of success in the 1990s.

Rewards for success can be expressed in many ways: money, prestige, management recognition, or promotion. However, it is very important to understand that people have different expectations and goals. The reward structure needs to accommodate individual motivations and goals.

Usually most sales people will say they are in sales to make money. Money is the prime reward expected. However, like anything else, people have different expectations concerning their rewards and personal goals concerning income. A salesperson working for a company may be "comfortable" making $65,000 per year, while another salesperson in the same company is comfortable at $85,000 per year. Other salespeople may seek to make $120,000 per year or more. Some salespeople expect to work only thirty to forty-five hours a week. While other salespeople expect a standard sixty-five hour work week. The individuals define the priorities in their lives and work accordingly.

I once told a manager of mine he needed to get someone else to penetrate a new territory. I did not know if I was willing to put in the sixty hours a week to get the job done. He said he wanted me because I could get more done in thirty hours a week than his other people could in sixty hours. I thanked him for the compliment.

Most sales managers appear to think everybody wants to make $140,000 per year and is working sixty or more hours per week. Not true. Each person has their own reward zone. What are the true income goals of your salespeople? Have you asked them? How many hours a week are they really investing toward their financial and career success? Are there disincentives in the compensation plan which make working longer hours non-rewarding or certain income levels unattainable in their eyes? Seek to understand the real goals and motivations of your salespeople so you are sure you are running the ship with full sails.

Aside from money, there are other creative ways to reward success. One company I worked for recognized and rewarded its successful peddlers through greater perquisites, such as a fancier title, larger car allowance, a nicer office, raises in base pay, bigger than average sales achievement plaques, letter(s) from the president, extra vacation days and so on. They made sure they recognized their successful salespeople without lessening their focus on their primary duty, which is to sell.

Also, a promotion can be considered a reward for success. However, often times promotions mean taking on new duties not related to sales or, God forbid, becoming a sales manager. This is okay if this job change is truly wanted by the individual and it is also in the company's best interest. But again, let the peddlers sell and let the administrators administrate. A promotion is not always the right answer.

Bonuses, commissions, and other rewards are in a sense paid for achievement "above and beyond the call of duty." The days of automatic pay raises and promotions are long gone. Unfortunately this is not what some salespeople have been allowed to expect. In too many companies, in spite of efforts to the contrary, today there is less relationship between success and financial reward, or failure and financial punishment.

For example, one company I am familiar with used such criteria as time in grade, internal references from non-sales functions within the company, and approval from two higher levels of management as their primary criteria to document readiness for promotion. The emphasis here was more on following the corporate culture than sales victories. Often a salesperson has to rock the boat and irritate management of non-sales functions in the company to win competitive business. How am I going to win more than my share of new business and get promoted when I get penalized if I rock the boat? Remember: Real sales contribution is effort which advances your company in the decision approval process and leads to successfully closing the order.

How is the rewards structure set up in your organization? Sales is a risk and reward environment. There is a carrot and a stick. If you do not see both where you work, things will change for you.

* * *

Commissions

It is interesting to watch computer companies with VARs, OEMs, distributors, systems integrators, and end-user sales forces attempt to develop equitable and administerable sales compensation plans. Obviously as more parties and locations get involved in a complex sales opportunity, it is easy to lose sight of who deserves which rewards. The rewards must go to the person(s) whose tangible contribution made the sale happen. Returning a telephone call or having lunch with someone does not automatically mean there was a contribution made to the sale. Simply having a company's name on your assigned account list when you have rarely been to the account does not mean you automatically deserve anything. Commissions are to be earned; they are not birthrights!

I remember when a commission check was a badge of honor...a competitor's wound. However, today some companies have gone so far as to hire internal compensation consultants to help explain to a salesperson what they can expect to be paid for a certain deal! Am I missing something here? No!

It seems many companies have gotten away from the basics. Limit the compensation plan's rules for payment of commissions or bonuses to a couple of pages of print, and pay in a timely manner.

Many compensation plans have rules in place to preadjudicate commission sharing or negotiations about commission splits which should normally occur between two or more salespeople concerning a specific deal. Sometimes these guidelines foster cooperation and teamwork, and sometimes they do not. If a salesperson avoids discussing splits in a straight forward manner with another salesperson, how successful are they going to be representing the company in a face-to-face negotiation with a customer? Split discussions are only unhealthy when three levels of management get involved and commission credit allocations become more important than customer satisfaction.

I worked for one company who paid more than one hundred percent commission for certain sales channels where the company wanted to foster greater cooperation between salespeople. These greater than 100% commissions rolled up to the branch manager, but not beyond. Also, any salesperson claiming commission credit on a deal who did not materially contribute to the success of the deal was subject to immediate termination by the vice president of sales. Therefore, few salespeople spent time chasing other salespeoples' successes. Finally, commission claims had to be made before the purchase order was received from the customer. It is funny how some people's memories change after an order is placed. Claiming credit after order placement caused management to wonder if the salesperson really was involved in the early stage of the deal.

Sales has traditionally been a high-risk, high-reward environment. Income was roughly fifty percent salary and fifty percent commission or bonus. I still believe that a commission check is a big carrot. Get back to the basics. Keep it simple.

* * *

Crisis in Confidence

The computer industry is not simply in turmoil today – it is in pandemonium. There are underlying trends which seem almost permanent:

- Constant organizational change in both vendors and customers operations – perpetual re-engineering
- Lack of stability on a personal and corporate level
- Declining profit margins
- Flatter organizations with reduced infrastructure support
- Devaluation of experience

These long term trends are further exacerbated by management's constant, ninety-day focus. Combine this with a work environment besieged by constant change and it is no wonder planning beyond an eighteen-month window becomes droll. Inconsistent quarterly revenues and declining profits have the stockholders and boards of directors continually breathing fire. Management tactics of consolidations, reorganizations, and layoffs will never end. It is now questionable if these long-utilized techniques for getting the house in order are as effective as they appeared to be. We are seeing them now too often.

All of these problems create a continuing and perpetual state of low morale. Senior management may not realize how much this affects an employee's productivity. People today have fear about constant change, uncertainty about management's agenda(s), and doubt about management's commitment to the economic well being of every employee. A person worried about their job, income, or getting a new boss next month is not focused on their work – Principles of Management 101!

Paradoxically, senior management sensitive to hearing evidence of low morale often tries harder to advertise an "open-door" policy. However, where are open-door policies real...sincere? More often employees encounter an environment where senior executives are hesitant for an employee to arrive unscheduled at their office and ask candid questions about the company, business conditions, or a recent management decision. Management needs to communicate sincerely, more often, more clearly, and listen carefully to all employees. Take off the insignia on the collar. Leave the ego at the door and listen. I do not care how we get there, but we must move ahead together.

Why do managers fear a subordinate having an impromptu chat with their senior management? Why are employees afraid to say what is really on their minds to the big boss? Why do workers have a lot of skepticism when they have an opportunity to talk with top management? Because they do not trust their management to make the right decisions – meaning what is best for the well being of the company and all of its workers, even though they rarely see the daily business issues faced by their top management. This crisis in confidence in senior management's motivations is the biggest problem facing companies today. Is top management focused on the long-term viability of the company or ensuring they receive an executive bonus next quarter?

As a result, the hi-tech industry has degenerated into a "hired gun" mentality. Have compensation plan, will travel. Building relationships and a long-term commitment are no longer a priority. This constant churning of sales, marketing, and business development talent is bad for everyone. There are no winners. The industry, companies, salespeople, and customers all lose in this environment.

Companies actively seek the most senior person, the person with the most knowledge and experience, but do not want to invest in the verbalized future strategy. In locating this new "guru," the company assumes college degrees, computer proficiency, and subject matter expertise are already among your skills. If you have all of these, you may get to go to the job interview. If you do not have all of these, good luck! Oh, by the way, having all of these assets does not necessarily mean an increase in pay. Today you can purchase computer software for your PC which offers expert-level, subject matter knowledge for $29.95. This combined with the glut of unemployed white-collar workers means that experience earned over many years no longer automatically commands an earning premium. The company is looking for a distinct skill — not a long term commitment. Instead of seeking employees in their fifties who have been there before, companies are hoping the mistakes of younger employees will not cost more than the savings generated by less payroll. This example of corporate America's ninety-day mentality clearly displays the trend of devaluation of work experience and the decrease in commitment to individuals.

Additionally, to leap ahead of the competition, companies bid big bucks for the latest "revenue savior," or consultant with the "vision," or super-duper technology. These approaches may help, but are they integrated with the company's discriminators, strengths and unique talents of its employees? I see many companies trying to bring in a "gunslinger to clean up the town." This hired-gun approach to jump start the company further reinforces the temporary nature of the company's commitment to the long term. Employment services are being bought and sold, and after twelve months no longer needed, or a new "strategic direction" is unearthed and off we go again.

The company wants loyal employees who will give that extra effort when needed. However, where is the commitment to the individual? Loyalty is a two-way street which many companies seem to have forgotten.

In turn, individuals want to work for a stable company with benefits that also pays well. Nobody wants to take any risks. When a company starts to enter a troubled phase, employees immediately start looking at their own career options. Where is the commitment to a company or getting a job done in good and bad times? A long-term commitment today seems to be nine months. This, of course, is another systemic problem for American business.

Customers are losing the most, because they are seeing a poorly trained, new face every nine to twelve months who tries to bring new levels of sincerity to the phrase "Trust me." How deeply will a customer get a salesperson involved in their strategic planning when their gut tells them that person will disappear at the end of the year? Some customers have arcane applications or technical environments it takes years to understand. Meanwhile technology platforms change every 18 months and new products are announced monthly. Who knows if the existing applications and new technology can be successfully married?

Right now all sides are losing. The work ethic this country was built on went beyond a 90-day-get-results-or-else mentality. High turnover in senior management, low dedication to a work ethic, and a lack of loyalty from all parties will continue until we stop this 90-day mentality fostered by investment bankers on Wall Street. A company wants to be successful in the marketplace. A salesperson wants to make a commitment beyond "What is the compensation plan?" A salesperson wants to believe in the products they sell and in their management. The customer needs quality products and services to achieve their evolving mission.

Let us have realistic expectations. We need to stop buying hired-guns – a person who is single-handedly going to double revenue, with little or no training or significant resources, within six months. TANSTAAFL! Go for the win-win, not the we-will-see-where-we-are-in-six-months approach.

*   *   *

You Too Can Be a Rocket Scientist!

One company I worked for had the philosophy that anybody could do almost any job given some training and enough time. This philosophy, of course, continued even while the company was experiencing significant layoffs. Can you imagine that? So, for example, people from manufacturing, finance, or engineering were allowed — even encouraged — to enter sales, regardless of their aptitude for sales. I always wanted to have the following conversation with the president of that company....

Bob: Mr. President, thanks for this opportunity to speak with you. I have been at Farble Corporation for six years in the sales organization and know layoffs are planned in my Division. I do not want to leave Farble and would like to do something new. I want to help design the next generation of microprocessors. This would be challenging and I promise to work hard.

Mr. President: Bob, I appreciate your enthusiasm and know you are a dedicated employee. But, Bob, you do not have the right background.

Bob: But, Mr. President, I will relocate; I am willing to go to the required six-week course offered by the company and, given some time, I am confident I will get the hang of it.

Mr. President: Bob, Bob, you do not understand. Microprocessor design requires advanced training, specialized knowledge, and, frankly, a talent for this type of work. Too much is at stake for Farble to view this as a simple cross-training situation. Trust me.

Bob: Well, why do you allow almost anyone in the corporation to enter sales? Sales requires advanced training, specialized knowledge, and a talent for this type of work, too!?

Sales is both an art and a science. Not everyone can be a salesperson or a sales manager. Sales requires not only specialized knowledge and training, but certain skills and talent. However, there are salespeople today who, though they do not have a propensity for sales, still hold the title and responsibility for sales. All of the training and time invested waiting for them to produce the needed level of play will never result in a positive return on investment. It will not be there. This is not a question of desire or tenacity. They are simply not equipped to do the job.

Many years ago I had the idea I would make a good basketball player. I was 6'2" tall and slender. I used to practice regularly, and frequently played in neighborhood pick-up games. In 1972 my basketball career peaked when I played on an intramural basketball team while in the U.S. Air Force stationed at Fort Lee AFS, Virginia. During our team's twelve-game season, I scored a total of two points – yes, two points. I slowly came to the realization I was not agile or coordinated enough to be a basketball player. I also did not seem to have the "mind" for quickly anticipating scoring opportunities while dribbling down the court during a game.

What if I had had serious aspirations of being a career basketball player? What if because of my height or large hands someone would have insisted I be a player on their basketball team? I could have practiced and practiced. I could have attended many basketball training camps. Maybe I could have even played some practice games with some professional players. However, no matter how much training could have been invested in me over a long period of time, I was not meant to be a basketball player. I do not have a propensity for playing basketball.

It is the same for some people in sales today. They are simply not meant to be salespeople. Yet, companies seem determined to reorganize, change quotas, schedule mandatory training, hire new sales managers, rewrite compensation plans, and so on, rather than accept this premise. Before you turn the company upside down in the pursuit of revenue, make sure your salespeople are truly peddlers with the skills and talent to be successful, and not administrators destined for mediocrity, at best.

* * *

Wanna be a Manager?

I once turned down an opportunity for a sales management position at a large company. The scenario was very similar to:

Big Boss: Bob, we have decided to ask you to coach our next company football team. We believe you will make a great coach!

Bob: This sounds great! How many players will I have on the team and when can I start building the team?

Big Boss: Fourteen players; and the next fourteen people we find, regardless of height and weight, will be on your team. Oh, by the way, you do not have authority to cut or trade any players. Isn't this opportunity exciting?

I decided not to take the job. They were looking for an *administrator*, not a sales manager.

* * *

Corporate Sales Training

The corporation has both a responsibility and a vested interest in training its workers. The standard training provided by corporations usually includes two major components: skills enhancement and product training. To be effective, all training must be crisp, focused, and updated regularly.

Corporations are faced with a serious dilemma in developing appropriate and current sales training materials for sales skills enhancement. The best person to train peddlers is another successful *peddler*, but a true *peddler* wants to sell and usually considers a job in sales training to be a demotion of sorts. As a result, the training course developers and instructors usually do not even come from a successful sales background of at least five years. Additionally, too often the training organization is considered an independent corporate resource. Consequently, the course developers and instructors have minimal interaction with the day-to-day issues facing their students. How can their training be successful?

To resolve this dilemma, some companies and individuals depend on "proven" internal sales training programs developed and carved in stone years ago. As technology is moving, so must the content and presentation of the sales training materials. Again, how can this training be successful?

The best resources to recommend improvements and additions to the sales training efforts are often the students themselves. Has the training organization incorporated changes to the curriculum suggested by peddlers or outside training consultants? Has the training organization even solicited feedback from outside sources? Take a fresh look at the training curriculum. Is it rigid? Is it presented as the correct Way? Sales skills training needs to be focused and current. The training organization needs to listen to the peddlers. Learning is a process that both students and instructors experience. The feedback loop to the instructor provides a wealth of information and knowledge that in turn can improve the process for all.

In addition to specific sales skills training for the sales organization, does the corporation recognize and capitalize on the sales potential throughout the enterprise? Sales skills training should not overlook other employees in the company who are capable of being extensions of your selling machine!

What are the company's discriminators? How do my company's products and services compare to the competition? A four-hour mandatory training course for all employees answering the above questions provides a solid baseline of understanding throughout the organization. For the future world of decentralized, empowered work teams, all employees can become selling resources, that is, an extension of the sales function, given appropriate information.

Product training must also be clear, crisp, and *relevant*. Product training for the "sales" audience really needs to provide information on two levels: technical specifications and specific marketing information. Good, targeted product training results in overall savings to the corporation. Lack of information in either arena puts the corporation and the salesperson at a disadvantage.

It seems the better product sales training I received, the less reliant I was upon the sales support staff. Sales support people are too valuable to take on every sales call. A salesperson should be able to make the first and second sales calls alone to qualify the account. He or she should at least be able to answer basic technical questions.

Product training that also clearly addresses the special needs of the "sales" audience in addition to the high-level technical specifications is valuable and respected. Some of the key issues the *peddler* wants addressed include:

- What data was used for market research in product development?
- What are the key discriminators and weaknesses?
- Why would my customer be interested?
- What business problems are solved by the product?
- How is the product positioned in the marketplace?

Few companies I am familiar with have outstanding *sales-oriented* product training. Having worked for several technology companies, I have attended a range of courses designed to equip me with relevant product knowledge intended to facilitate my sales success. By and large these attempts at product training have been a break in the day-to-day routine at best and totally unproductive for course attendees at worst. Having worked for small companies with limited internal training resources, I have had to learn to train myself – by reading a brochure, reviewing a technical manual, or by pulling teeth from the product manager. I would then organize this material into a presentation of twenty overheads and go forth to spread the word to uninitiated customers.

Many years ago I hesitantly attended a front-end processing, communications products sales training class at Tandem Computers. What a pleasant surprise! Instead of the usual smorgasbord approach to training of "here are all the technical details; you figure out what is important and go tell someone," I received real product sales training! The agenda included in this general order:

- Product Engineering - two hours - a real product developer presented the product specifications they had received from marketing; why the product was designed the way it was; which features were included, will be included or were ignored and why, from engineering's perspective; why the product was the size, shape, color, weight, and so on, it happened to be; why engineering believes the product is unique; challenges encountered while designing and developing the product.
- Product Management - four hours - the actual product manager presented why they gave engineering

the specifications they did; sources of their market research; how the product is used in business(es) – the problems it solves; key features and benefits; whom the competition will be and their strengths and weaknesses; product performance data; how pricing was determined; and a forty-minute standard sales presentation of the product as if the product manager was presenting the product to a new customer; and what to expect over the next four months in product changes or enhancements.

- Product Demonstration - one hour - showed us obvious and subtle features of the product appropriate for a customer demonstration.
- Customer Testimonial - one hour – two executives from two companies in separate industries who had just purchased this product gave candid comments about what they liked and disliked about the product; why they selected it over a competitor's offering; and what they expect a salesperson to say to them, if you expect them to buy another one.
- Salesperson Testimonial - one hour - a Tandem salesperson presented in some detail a recent sales victory where the product was key to a new account win and why.
- Sales Training Person - one hour - presented all marketing collateral material available; gave out standard overhead presentation material, and a list of resources and contacts to support selling this product during the sales cycle.
- Closing Presentations - one hour - two or three class attendees were chosen at random by the training person to give a ten-minute spontaneous presentation

> of the product using any of the materials at hand. (You had about one hour to get ready to present to the class.) Knowing this presentation was coming kept people listening and avoided people popping in and out of the class to take care of other activities.

This is product sales training at its finest! When I left this class, I understood and knew how to present Tandem's approach to front-end communications processing. I still remember this training class today.

It was nice to work for a company that was serious about the quality and role of their sales training program! Tandem's training policies and procedures further demonstrated their focus on training and continuous improvement. For example, you could not pass this training class simply by initialing the roster. You actually had to attend the class and pay attention, instead of simply making guest appearances during the day, as is often done in these courses. You could not get official credit for taking the course unless you personally gave a completed critique form to the instructor. The instructor, upon receipt of the critique, gave you a signed certificate which you had to present to Human resources for entry in your personnel file, so you could receive formal credit for taking the course. At performance appraisal time, your training record was actually reviewed to ensure you were taking and progressing through specific (technical and non-technical) training courses, appropriate for your career path.

These positive examples provide a clear model for a successful corporate sales training program. Take a fresh look at your training curriculum. How do you measure up? Is your organization using the latest tools and information available or have they all together said "No!" to outside training methodologies? Beware! Sales lives in a competitive environment. The individuals with the best, most current information have the competitive edge. Is your sales organization being given the edge they need?

* * *

Commercial versus Government Sales

Few large commercial companies truly understand the complexities and vagaries of selling to the federal government. However, because once upon a time a member of the senior management team sold nineteen-inch equipment racks to an Air Force base in North Dakota, the company feels equipped to play in the federal procurement arena. This is a dangerous assumption.

When dealing with the federal government, it is imperative to understand some basic differences. There are differences in accepted business practices. The Procurement Integrity Act of 1989 requires a higher degree of ethical conduct than commercial sales when selling to the federal government and its procurement officials. There are questions concerning funding or evaluation criteria, for example, you simply cannot ask the federal government customer in some situations. This is forbidden by federal law and in fact has stiff penalties for violations of this law. Has your upper management reviewed the spirit and intent of this law?

There are differences between the Uniform Commercial Code and the Federal Acquisition Regulations. There are differences between the company's fiscal year and the government's proclivity for buying sporadically over a period of years on the same contract. There are differences between short-term product business and *program* business.

Occasionally a commercial salesperson newly assigned to a major federal account, who has none of the requisite federal government procurement training to sell into this complex market, will generate a high revenue quarter. The home office is very pleased and thinks he or she is doing a fine job! They also think all that whining about government sales being "different" has just been proved wrong. They do not realize that the spike in that quarter was the result of work performed far back in the sales cycle and had no relationship to the individual currently assigned to the account. This potential revenue never made it on the ninety-day forecast. Management often has no clue as to the amount of lost revenue in the federal programs arena. Why? Because they never see it on a forecast and are rarely told about lost opportunities.

It is clear that the impact of these differences is not understood at the corporate level when computer companies give their federal government salespeople a commercially-oriented compensation plan. Why do companies throw all of their salespeople into one bag with one compensation plan, especially unfair to their federal salespeople? Companies talk about taking federal business seriously and yet take for granted their salespeople who call in the most complex and challenging technology marketplace of all.

These commercially-oriented compensation plans penalize the salespeople with government accounts because sales measurements and bonus programs are usually tied to short-term time frames, e.g., ninety days which have no meaning to a long-term government customer's buying plans. Year after year, the so-called incentive plan causes frustration, turnover, and missed forecasts, because the commercially-oriented plan is unfair in the government environment and, in fact, is a large disincentive to salespeople with federal government accounts.

Federal salespeople with other things being equal should get a higher base salary, for example, $10K more than their commercial counterparts. They may receive a smaller commission percentage, since their orders can be rather large. They might be measured on a twenty-four or thirty month basis, instead of the standard twelve months like the commercial salespeople. The government often engages in contracts lasting five, seven, or ten years. Why measure a salesperson's success in a single year in the middle of a long-term contract? It often takes the federal government two years after contract award before volume purchasing begins. Additionally the federal salespeople should have ongoing training to make sure they understand the federal procurement environment and its plethora of unique buying "channels."

Professional people know what they know and what they do not know. Who is managing and driving your federal government business strategy?

I Guess They Don't Want My Business.

I answered my telephone early one morning with my usual, "Bob Davis." My friend Michael said in an excited voice, "I can't believe it. After two and one half years of laying a foundation for the new project's computers, they pulled him off the account!"

"Who pulled whom off the account?"

"They pulled Ron Yankoupe off our account. He has done a great job helping us lay out the new architecture. He is technically competent and has been responsive to all my requests. Ron has worked long hours, way beyond any other salesperson on this project."

"Michael, you know Ron's company is laying off many people. They can't keep having their salespeople doing free consulting. There has to be some tangible revenue on the horizon."

"Bob, are you crazy? We plan to buy over $20 million worth of computers early next year. This purchase is at risk for his company now due to Ron's leaving the project. I must have regular access to him to finish the job we started. You wouldn't believe how two of Ron's competitors are outside my office each week begging for a chance to bid these systems. I can't believe Ron's management is so nearsighted."

"Michael, Ron's management has been hearing about these orders for over two years.

They have to cut costs. If they can't see and touch an order today, they do not care about next year! They won't be here next year!"

"I hear you. They will never know how much business was theirs for the taking. They must be in more serious financial trouble than even I realized."

"Michael, if you feel so strongly about this, why don't you call Ron's management?"

"No, Bob. If they are this dumb, I will not help them. Besides, I am supposed to treat all of the vendors fairly and I guess the other company's vigilance is going to pay off."

Davis' Rule Nine

Sometimes companies have too much vision....

Vision or Strategy or...

I have been to many meetings where the "V" word – vision, is discussed or pondered. Early in my career I believed vision in a corporate environment related to direction and leadership. It was something formed from ingredients including analysis, logic, skills, resources, investment, competitiveness, and leadership. The process of moving from vision to strategic planning to budget development and consolidation to tactical planning to communication with all employees to execution to measurement and reward is logical to me. However, vision, as currently articulated, no longer seems to drive or encase the process. Vision has itself become a buzzword, not a rallying point.

True, I am not invited to Board of Directors' meetings or quarterly executive reviews. I am told vision is presented and reviewed at these internal meetings. Occasionally I receive a memorandum or pep talk from headquarters where the "V" word is mentioned.

I consider myself a mid-level employee who frequently meets with senior customer management. My company's vision and foci should be clear to me and ready to be shared by me on a moment's notice. In recent years I, as a *peddler*, have not been able to articulate my company's respective vision to our accounts. I have opinions and ideas based upon the memorandums and brief conversations with senior management, but I am not really sure. So like any good *peddler*, I develop three overheads based upon my perception and go forth to tell the story.

In an industry which foments and avoids change at the same time, wrong vision or uncommunicated vision is deadly. If you do not clearly express your vision both internally and externally in an organization, then it will be done for you by customers and, even worse, competitors. Management is so caught up in quarterly numbers and the price of the stock; how can they realistically envision the marketplace in 1998?

One company I worked for had its annual strategic planning retreat with charming facilitators where great ideas were listed on an easel. Vision was discussed and debated. Someone took notes of the many ideas. Soon the retreat ended and the realities of day-to-day business overcame the time needed to refine and improve the possibilities envisioned at the retreat. I am told the company's president is now concerned that the upcoming executive retreat needs to be better focused... .

The welfare and success of all employees, and in turn the company and the stockholders, demands this commonplace scenario end. Companies today need less of the "V" word and <u>clear</u> communication of the vision being espoused behind carved oak doors. Let's get beyond the words and get to the execution.

*   *   *

Industrends

There are trends underway in the computer industry that must be identified and understood by management. These computer industry trends are called industrends. Industrends are defined as long-term economic or marketing trends that affect a specific industry. They can be positive or negative. Today, they are directly affecting the long-term success or failure of hi-tech companies in our industry. Some of these trends are obvious and are putting companies out of business. They are offered not as a revelation or panacea but a basis for dialogue – dialogue between top management, middle management, marketing, engineering, and, most importantly, customers.

• Products are not Boxes!

Top management in hi-tech companies must understand what a product is and what it includes. A product is not just hardware, software, and documentation. A product is:

- hardware
- spares
- diagnostics
- advertising
- manufacturing investment
- profit margin
- warranties
- logistics integration (into existing environments)
- competitive analyses

- new selling approaches
- software
- user-friendly documentation (in a variety of media)
- sales and support staff training
- customer training
- public relations
- marketing collateral
- customer support plan
- compensation plans

A well-developed product plan ties all of these variables together. The time and financial resources necessary today to bring a successful product to market cannot be recovered if a faulty decision is made. The marketplace is most unforgiving today compared to the 1980s. Top management must resist making hasty decisions after a slick R&D presentation from some well-meaning engineers.

• Customers do not buy "Hot Boxes."

Users of computers today are evaluating enterprise-wide networks, multiple (dissimilar) on-line data bases, and competitive issues to support their organization's information requirements as never before. The ability to simply sell a "hot box" and be profitable is gone. You may think this is obvious? A lot of companies are still trying to sell hot boxes. As organizations make careful evaluations of their capital investment decisions the length of the sales cycle can get longer and longer. Furthermore, as time goes by, market requirements will change. Therefore, the vendor's product plan becomes antiquated halfway through the product's intended life. Most companies today cannot adapt to this rapid pace in the marketplace. Systems and products today from the same vendor are being announced monthly and replaced in fifteen months. Is your company in this race at all, let alone positioned to win?

• "Mom and Pop" companies are going out of business.

There are two kinds of highly successful companies today – the one-stop-we-have-it-all and the very specialized we-do-one-thing-best company. You see this today in restaurants, grocery stores, and the computer industry. Successful restaurants today offer quality, service and a broad menu, or the "world's best" corned beef sandwich, or cinnamon roll and so on. Grocery stores today are going the route of Giant Food and Safeway Stores or specialized gourmet shops. The mom-and-pop grocery stores are becoming rare. Hewlett Packard Corporation and IBM Corporation, for example, offer a wide range of products and services. Other successful hi-tech companies are focusing on a market niche and being the best in that niche – best in class. The resources necessary to be "all things to all people" are not available to most companies today even though some management teams have trouble admitting this.

• Standards are here to stay.

The promulgation and acceptance of standards forces a "me too" image or perceived commonality in vendor's products. The challenge is that even though standards may let everyone "connect" to the system, the cost of ensuring "connectability" goes up for the vendor and down for the user. Vendors used to be able to develop and market proprietary (real or perceived) product advantages and features. Today unique product features or benefits useful in a proprietary setting are quickly vanishing. Making the product faster, cheaper, and compliant with standards seems to be the entire focus of product development. This dramatically increases the pressure on the marketing function. The marketing function must become very sophisticated; marketing must lead companies through this challenge of selling services or value added versus selling perceived commodities. With ever diminishing margins, selling commodities or boxes, as I like to call them, has a grim eventuality. Services, for the present time, have enduring margins when compared to boxes. Marketing must lead a transition from boxes to services to expert knowledge.

- **Products and Services Will Become Commodities.**

Diagram B, Davis' Commoditization Curve, illustrates the trend of the Commoditization of the Hi-Tech Industry. Time and competition have slowly advanced products and services along the growth path to become commodities widely available in the marketplace. It is the job of marketing and sales to slow down this progression to make sure true or perceived product discriminators, durable competitive advantages, and unique product values are recognized by the customer. If this task is not done by marketing and sales, then our value to a business transaction is nil and we will be replaced by an online product catalogue and a toll-free order line.

Davis' Commoditization Curve - The Commoditization of Hi-Tech Industry

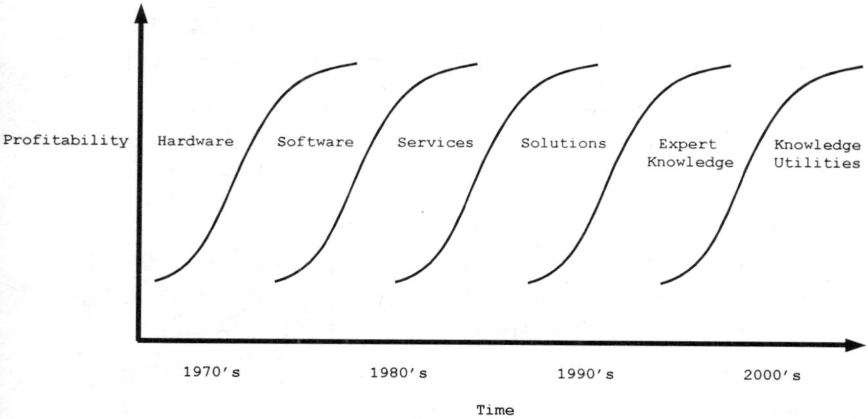

Diagram B

Commoditization of Hi-Tech Industry

There is a trend in our industry which has been underway for many years – commoditization. Commoditization is the process whereby all products and services become available at an ever lower price requiring less and less support and maintenance from the developer or manufacturer. Therefore the probability of additional revenue and profit after the initial sale lessens each year. Additionally, the methods used to market and sell commodity products include indirect channels and direct phone lines, not an expensive end-user sales force.

As presented in *Diagram B*, Davis' Commoditization Curve, commoditization has engulfed hardware and software, and is now engulfing services. Progressively, the commoditization curve establishes a "food chain" of perceived value and service. Hardware is at the bottom of the food chain and represents those vendors selling hot boxes, parts, and components. Software consists of commercial off-the-shelf products requiring less than 10% customization or tuning. Services comprises providing specific hi-tech skills that can address common industry problems. Further up the food chain, solutions represent those people selling tailored combinations of hardware and software designed for an application environment. Expert knowledge is sold providing expert-level skill focused on solving a multi-dicipline, complex problem. Last, and at the top of the food chain are information utilities. These are combinations of expert knowledge and solutions combined into repositories which are enterprise wide in terms of accessibility. These information warehouses or knowledge sources are available to all customers or users on an as-needed basis to facilitate their decision-making process within their industry. Executives will be able to soon enter known and estimated variables concerning a complex business problem and have several alternative courses of action quickly at their desktop. The pros and cons for each course of action will be presented clearly after including consideration of all risk factors. This will facilitate real-time tactical decision making in the trenches. The packaging and pricing of these knowledge sources will parallel the utility industry, i.e., pay for what you use, when you use it, where you need it, without regard to underlying technology or infrastructure. The notion of knowledge utilities is in its nascent stages but is definitely on its way.

Corporations will pay a premium for perceived value added and quality. Services and knowledge designed to promote the success of organizations while keeping underlying technology transparent will command higher margins.

The professional hi-tech salesperson must aspire to work in these environments and markets or technology itself will eventually replace or reduce their added value to the process of promoting successful commerce. Early in my career I sold "bits and bytes," or "speeds and feeds," and made a reasonable income. Later, as a business development manager, my role is to sell services and expertise to help my customers anticipate market opportunities. For my value to the corporation to increase, I must now also be able to show customers I can bring expertise to bear to solve their business problems. Additionally, job opportunities I have seen recently offer the most reward when I have been able to demonstrate how my existing business relationships may facilitate the success of the organization.

Bottom line: the only way a hi-tech salesperson can avoid having his or her job replaced by a catalogue and a 1-800-HOT-DEAL rapid order line is to move up the food chain. Move away from selling boxes to selling solutions and expert knowledge.

* * *

The Marketing Progression

There has been an overriding marketing progression for our industry since its beginning.

Concepts Ahead of Actual Products!

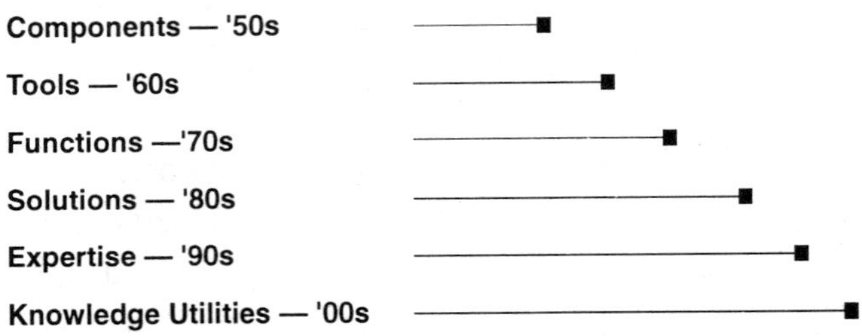

Diagram C

This trend represents how marketing concepts have led actual products and services since the 1950s. Today companies want to acquire expertise, that is expert knowledge applied to a specific business environment faced by a firm. The increasing volume of outsourcing today is a simple example of this trend. The president of a company in Virginia told me his firm outsources their strategic business planning. This fascinated me, so when I asked why, he said, "Because we are so busy with day-to-day matters it would not get done properly. Second, we have too many biases to look at our strengths and weaknesses objectively."

Taking this marketing continuum a step further, the marketing theme of the year 2000 and thereafter will be knowledge utilities. I mean knowledge, not information. Modest information utilities exist now, and are successful. America Online and CompuServe are examples of information utilities. A knowledge utility will combine a wide range of information sources and technologies with expert-level, business knowledge to be able to rapidly address and solve business problems approaching real time. This knowledge service may be done on-line, interactively with business executives as the need arises or as an ancillary role in a variety of strategic and tactical planning settings.

Clients will be charged based upon the volume and the nature (profile) of the knowledge transaction. There will be a range of values associated with certain expertise, which will in turn determine the amount of the monthly knowledge-utility Bill. This trend will cause a powerful shift in the role and value of information technology management in corporate America. Chief information officers will truly become chief! It will also directly affect the current level of investment in the computer systems scattered around the building. Ultimately all managers will have to be in the front lines – at the "edge of the market" or technology will not only devalue their experience, but obviate it.

Manufacturers developing hi-tech products will have to understand how their products fit into the world of knowledge utilities. How may their products support complex real-time problem solving? decision modeling? business simulations? neural network-based storage systems? AI-based competitive analyses?

Hi-tech corporations must also move their product offerings up this "food chain". For long term survival, their marketing focus must pursue the marketing progression noted earlier! *Diagram D* shows the primary marketing foci and *Diagram E* shows the future environment as customers move toward buying expertise.

Primary Marketing Foci

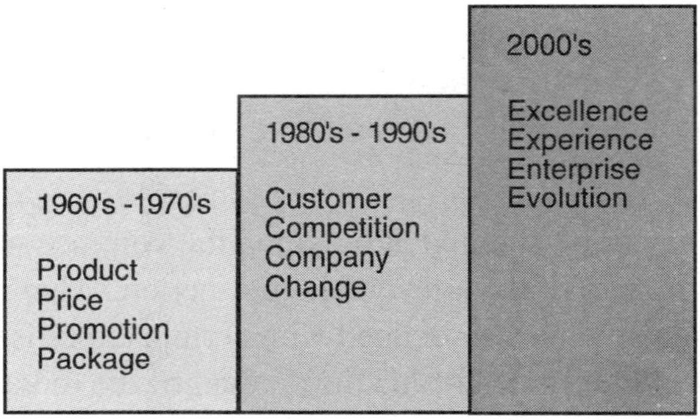

Diagram D

The conceptual marketing progression goes hand in hand with the shift in primary marketing foci as cited above. The four Ps of the '60s and '70s – product, price, promotion, and package were superseded in the '80s by the four Cs – customer, competition, company, and change. We are just now beginning to see a shift to the four Es – excellence, expertise, enterprise, and evolution. This will be necessary due to the even faster changing nature of the hi-tech marketplace.

Excellence will entail being the best-in-class of what you do. TQM, ISO 9000, and Carnegie-Mellon's Software Engineering Institute's (software process) Capability Maturity Model (CMM) assessments will become minimum standards. Corporations will strive to become the standard envied and respected by all competitors.

Expertise refers to the level of expert knowledge in your firm focused on specific application environments. This entails the ability to apply leading-edge technology to complex business problems.

Enterprise represents the broadest meaning of the term and will mean boundaries for a firm which from a management and technology perspective will include the environment, customers, suppliers, consumers, social issues, and the government.

Last, evolution involves capitalizing on constant change in the nature and scope of products, services, and skills offered by an ever-changing organization. It is the anticipation of and understanding how time itself and technological change will obviate the current marketing plan. Will this change affect my business plan in four months or fourteen months? What are my alternatives, countermoves, and so on?

Understand the four Es for your organization now! Look at the marketing function in its broadest possible context. Begin the move from vision to action now – before you are left behind.

* * *

Future Environment for IT Marketplace

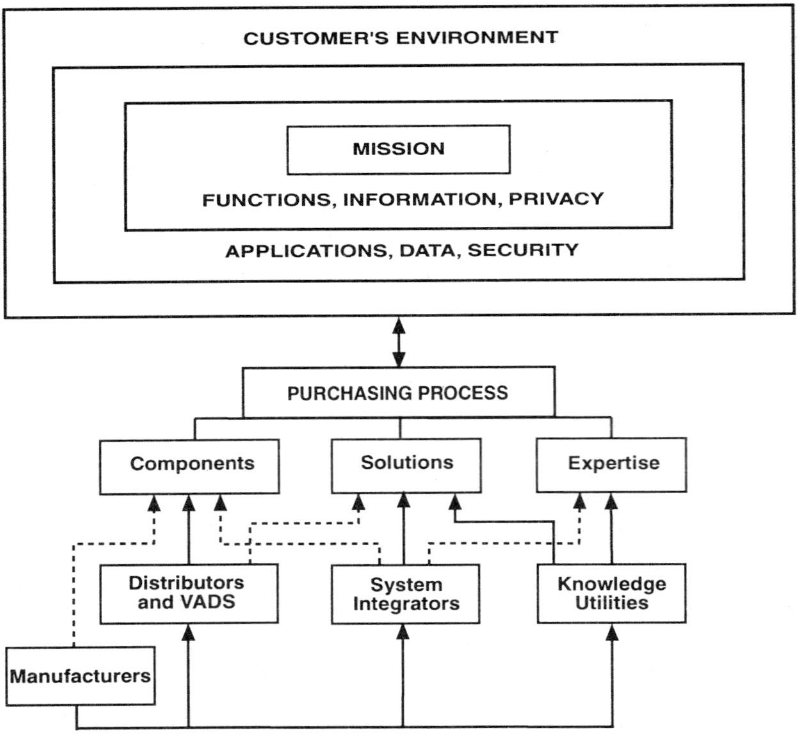

Diagram E

Your Turn!

I was going to write a vignette here describing a conversation concerning vision that I had with the president of a company in Virginia. However, I decided it would be more beneficial for you, the reader, to take a short test. Your assignment is to write in the space below within two sentences your company's vision. You have five minutes to complete this task. Begin now.

Davis' Rule Ten

Professionals find and use their hidden assets.

Hidden Assets

All of us have two assets which are frequently under utilized or hidden: creativity and the power of thought. Salespeople who are successful year end and year out in a competitive environment have learned to access and utilize these assets to their benefit.

All people have creative talents even though they may appear hidden. They have intuition and insight available to help them make the right decisions. New ideas and strategies come from this "creativity locus." We have to learn how to access this internal resource. To tap into this resource, we must learn how to listen or engage in the art of inner listening. If we are emotional, tired, or overwhelmed by events, the door to our creativity locus is virtually closed. If we are calm, quiet, and can clear our minds and listen, simply listen, wisdom and insight will come forward. Meditation or contemplation are practices, among others, which can help us learn to calm our minds. This process takes practice, as all worthy things do, but it works to release the creativity inherent in us all.

Thought is the other powerful gift. Thought has a power unto itself. Powerful thoughts are contagious; they infect people. Positive thoughts create positive results. If a salesperson strongly believes a win is not only possible but will happen, their leadership will infect the rest of the organization. Their belief becomes a self-fulfilling prophesy. If they go to an internal meeting and talk teamwork and winning, but do not believe this 101%, it will be sensed by other employees. They, in turn, may develop negative thoughts which, in turn, can undermine the success of the entire deal. Just do not say something, believe it! Use the power of thoughts to your advantage.

We are individuals. We each have a unique blend of talent, skills, and experiences. There are an infinitesimal number of routes to the goal line. We get to choose our own path across the field. Sometimes we forget to trust our own judgment, to accept our own innate wisdom. You must learn to listen to your instincts or as Howard Thurman once told an audience: "...you will all of your life spend your days on the ends of strings that somebody else pulls." This process is called "hearing the genuine in yourself," that is listening and hearing your individuality. Listen to your inner self! When you gain new insight or an idea to increase your success, grab it. Use your own gifts! A professional finds and uses his or her hidden assets.

Making Choices: One Salesperson's Priorities

I noticed Dan, a senior salesperson who worked for me, would quietly leave the office for a few hours during the day, about twice a week. His absence was sudden and secretive in nature. Dan was a popular person in the office, but struggled to make his numbers. I had spent time with him helping him prospect for business and gave him extra leads to help him generate new business. Still a revenue stream never developed.

Finally one day I called Dan into my office for a chat about his mysterious disappearances. Dan calmly explained to me he was an active sponsor in Alcoholics Anonymous (AA). He had had a serious drinking problem himself which cost him dearly before he had joined AA years ago. He said he owed a lot to AA and had to be available to help others. He "truly knew their pain."

> I said little and did nothing. I decided to leave him alone. I felt instinctively the help he was providing to others in need was important; important in a broad sense. He knew the professional versus personal trade-offs we all must consider. There was no need for me to say more. He continued to work with AA members on a regular basis. Dan also continued to struggle to close new business and did not achieve his quota.

Making a Contribution

Like everyone, I realize I will live and die after living a life which was the result of my own decisions. No one wants to leave this world without making a mark – a contribution. All of us want to be remembered. We want to believe our lives had significance long after the last heartbeat.

I have written a few articles, served as president of a professional society, donated time to help others, given money to churches and charities over the years, and given numerous presentations to share my industry experience. I hope I have had a positive hand, albeit less than I liked, in the growth and development of my two sons. Still, we strive to leave a bigger mark.

Can success in our jobs be a long-term contribution? Our hi-tech sales success promotes economic growth and security through new jobs and business expansion. It was once said that every new computer sold created about 40 new jobs. Is this not a mark? Maybe, maybe not. The application of new technology can foster layoffs and consolidations. Will we be remembered as givers or as takers when we performed our job duties?

An Unusual Application

One day my inside salesperson told me a small division in his company was considering buying a computer so they could stop renting processing time on a remote host computer system. This division was called the Armaments Division and I did not know the nature of their work. I did know a dedicated small computer could save them a lot of money compared to paying for timesharing on a remote mainframe.

I called and made an appointment to see one of their technical managers who frequently used the remote processing service. This person was polite and knew his company had a long-term, successful relationship with my company. I arrived at the designated time and the meeting began in a cordial fashion.

The customer immediately began to explain the high costs of doing simulation modeling on the remote mainframe computer. They needed to get a small, in-house system. Our computer was in fact designed for FORTRAN simulation modeling applications and the customer was entitled to a significant discount due to the volume of equipment purchased by his company each year from my company. This was going to be an easy purchase for him to justify.

I then asked the customer what they did in his department. He said they designed and tested anti-personnel devices. I asked him to explain this further. He said they were trying to design a new type of cluster bomb, dropped by an airplane, which would detonate about five feet off the ground, spraying plastic needles in all directions. Plastic was being tested because it was difficult to locate once it entered the body. The customer added, they had placed balloons four to six feet above the ground across a football-field-sized area and began live testing, but they needed to use the computer simulation application program to improve the design of the device to make sure they were maximizing the kill ratio. Did I have any questions?

I sat quietly and said nothing. I was stunned! I chatted with the customer briefly then left the building. I never called this customer again or pursued this opportunity. I never discovered if they purchased an in-house computer system.

I am fully aware computers can be and are used for purposes of promoting death and destruction. However, this conversation upset me. I thought, "No!" I will not sell computers to help maximize kill ratios. Was this right? Did I have the choice of saying no when employees in manufacturing were desperately trying to hold onto their jobs? Was I forcing my moral standards on others? Is this the right way to make a contribution?

After many years of working and pondering this question, how to leave a mark, I believe work is the means, but not the end. There must be balance in our lives. There is positive and negative in most things, and too much time invested in one activity has a cost. My parents told me when I was 14 years old, they did not care what I did for a living when I grew up as long as I was the best.

They also told me to make sure I did not forget my obligations. Our responsibilities include family, friends, co-workers, customers, and total strangers. It is our choice to help, to extend a hand or put off personal involvement until we have more time....But what is the cost of our delay? What rewards have we failed to grasp?

If you simply get up and go to work, then at the end of the day go home, then get up and go to work, week in and week out, you are a one-dimensional person. The one dimension being work! Some day you may begin to wonder if there is more to life than working and paying bills. When these thoughts occur, they are a subtle wake-up call, a call to try a new path. Yes, there is more to life. Life has other dimensions. Sure, you are aware of this, but what do you do about it?

Our work and job duties entwine us in the lives of our co-workers and customers. Despite efforts to keep our private lives separate from work, each of us gets glimpses or even involvement in other's lives. We see the funny and the sad. Our comments or counsel can influence the actions and thoughts of those around us, even when we are not aware of this. All of us want approval, to be accepted, to be remembered.

Each of us has his or her own opportunities to see the roses. For some it is involvement in church, for others a spiritual journey, donating time to charitable activities, or spending more time showing love to family, friends, or strangers. These other dimensions of life are important – crucial. They give balance to our lives. They enrich our well being and help us see our higher self. This, in turn, will increase our success as business people. Time spent giving, loving, and sharing with others is our greatest investment. Move beyond one dimension....

All In a Day's Work

I was getting ready to leave Loring's office when he asked me to go to lunch. I had just stopped by his office to quickly drop off some technical manuals and check on the status of the computer system. Loring was an electronics engineer using our computer system to control cockpit instrumentation for an F-16 flight simulator. It was critical that our system execute its software in a timely manner.

I was in a hurry, but decided I had time for a sandwich at a nearby restaurant. Loring and I went to the restaurant and ordered quickly – a tuna melt and iced tea for me and a Scotch on the rocks for him. Immediately I realized Loring wanted to get away from work to talk. Loring slowly began to talk about his frustrations with his work and management. He continued to talk over drinks while I sipped iced tea. I wanted to leave, but knew he wanted to stay.

While half-listening to Loring I began to look around the dining area and realized we were the only people there except one other table. Seated at this table was a family of six, apparently there to celebrate their grandmother's birthday. I thought this was special and noticed the elderly lady was enjoying herself with a beaming smile.

Loring suddenly called my name, "Bob, Bob?"

"Yes."

"Answer me. Do you think I should accept the job offer in Florida?"

I realized I had been ignoring Loring and began to conjure up an answer like, "The grass always looks greener..." when there was a loud crash as a chair fell over at the table with the family. I turned my head quickly to look. A man had jumped up and was panicky. He was pointing to the elderly mother who was apparently choking and could not breathe.

I watched for a few seconds and then got up quickly. I walked up behind the woman and immediately began the Heimlich Maneuver. After about four hard squeezes something in her throat cleared. She began to gasp for air and just sagged into her chair when I let her go. After a couple of minutes she seemed okay; so I went back to my table.

Loring was pensive. He asked me if I wanted more iced tea. We sat there silently sipping our drinks. The man at the other table came over and thanked me. I said little. Shortly, Loring and I got up and went back to work. I had to prepare for a meeting, early the next morning, with a key customer. Such is the life of a *peddler*.

* * *

Appendices

Appendix A

Ordering Information

You are invited to share your comments or questions. We will respond to inquiries concerning integrated sales force automated systems, professional sales and sales management training. Please write to:

World of Hi-Tech Sales
1153 Rt. 3 North, Suite 84
Gambrills, MD 21054
410-787-2492

The price of the book is $20.00 which includes applicable taxes. There is a 10% discount on orders of six or more books. Educational and non-profit organizations are entitled to a 15% discount if six or more books are ordered concurrently. Checks or money orders should be made payable to: World of Hi-Tech Sales.

Corporate Purchase Orders **ONLY** for six or more books may be sent by facsimile to 410-721-6494. Individuals, please send your order with a check or money order, by mail, to the above address. Please identify a point of contact and telephone number on the Corporate Purchase Order Form. For shipping and handling add the correct amount based upon the total quantity of books from the following chart:

Quantity of Books	Total Including Shipping & Handling
1	$3.50
2-5	$6.00
6-10	$10.00
11-20	$15.00
21-40	$25.00
41-75	$50.00

Appendix B

Glossary

Administrator - An intelligent, well-meaning person who adds no value to the selling process.

Big win - A competitive deal won when management said it could not be done.

Boss - A person who has the position of manager but is not a leader.

Box - A distinct hardware or software, off-the-shelf product usually sold in large quantities requiring little or no support.

Chalk talk - The use of a whiteboard or easel to get the customer to diagram his or her environment and how your products or services fit into their system.

Coffee call - A customer call which appears informal even social, but in fact has clear objectives to get specific information.

Commodization - A process whereby all products and services become available at an ever lower price while at the same time requiring less support from the developer or manufacturer, thereby generating diminishing profits.

Draw - Monies advanced to a salesperson, on a recoverable or non-recoverable basis, in anticipation of earning sales commissions in the future.

Face time - Time spent one-on-one with a current or prospective customer discussing business.

Follow-on business - Orders received from a customer after the initial order.

Industrends - Long-term economic or market trends that affect a specific industry in a positive or negative manner.

Inside salesperson - A customer who will go out of his or her way to help you win business in their organization.

Lay-off paralysis - A state of mind entered by employees who fear taking any action which may be viewed unfavorably by anyone in management.

Mindshare - When a customer has a business problem, they immediately think of your company first to find the solution.

Peddler - A person who was born to sell; a salesperson's salesperson.

Program - A specific business opportunity lasting a finite period of time, usually longer than one year with a variety of products and services which require some form of integration.

Psychological distance - Where a personal relationship is established with an individual customer while always maintaining integrity, respect, and discretion.

Qualified business - A sales opportunity where a company has a better than even chance of winning the business (relative to the competition) even after all normal qualifying variables are considered, such as funding, management relationship, pricing, ability to perform and so on.

Qualifying - A process to ascertain the prospect's needs, wants, and budget, and therefore be able to decide if you should continue to pursue this opportunity.

Revenue savior - A person who will single handedly cause a large increase in revenue in less than six months of being hired.

Reward zone - The level of compensation at which an employee is satisfied regardless of their job duties or performance.

Sales team - Any employee be it salespeople, support staff, administrators, secretaries, receptionists, telephone operators, mailroom clerks, manufacturing employees and so on, who are in a position to contribute directly or indirectly to a sales victory.

Success linkage - Where separate employees, departments or functions in an organization are tied together in a positive way to foster mutual success in achieving similar goals in a specified time frame.

Sugar coat - Making something negative or inappropriate, sound or look better than it is.

Synergy - That state of being sought by management trying to get a difficult job done with no additional investment of resources on their part.

Weenies - Director level or higher in an organization. Super weenies are senior vice president level or higher.

White paper - A general document, usually less than twenty pages including graphics, which presents an approach, concept or solution to address an account or industry's business problem from a management perspective.

Appendix C

Acknowledgements

Strategic Management Skills by Daniel Power, Martin Gannon, Michael McGinnis, David Schweiger, 1986, Addison-Wesley Publishing Co., 270pp.

Developing Business Strategies by David Aaker, 1992, John Wiley & Sons, 394pp.

The Art of Inner Listening by Jessie Crum, 1975, Quest Books, The Theosophical Publishing House, 96pp.

The Random House College Dictionary, 1984, Jess Stein, Editor in Chief, page 1312.

Eden's Guide by Trevor Mountaine, 1993, Heyberg Media, 242pp.

"When Growth Is More Than Numbers," March 13, 1994; sermon delivered by Rev. Dr. Fredric Muir, Annapolis, Maryland.

Sales Force Systems, International, 7790 S. W. 145th Street, Miami, Florida 33158.

Appendix D

Colleagues

Kathleen Coyle, software development executive, Massachusetts

John Gorman, financial guru, Maryland

Gary Hegna, company president, Florida

Al Hoffman, sales training guru, Florida

Chuck Hough, technical guru, Virginia

Neil Humphreys, professor of management, Virginia

Tim Karney, publisher, Virginia

Bob Lam, *peddler*, Virginia

George McFarland, sales executive, Florida

Terry Miller, federal procurement guru, Virginia

Margaret Mix, company president, Maryland

Dawn Placek, *peddler*, Maryland

Chuck Pross, computer industry guru, Maryland

Mike Wilmer, *peddler*, Virginia

Ruth Yankoupe, marketing executive, Virginia

Appendix E

List of Acronyms and Abbreviations

AI	Artificial Intelligence
CMM	Capability Maturity Model
DMP	Decision-making Process
I/O	Input/Output
ISO	International Organization for Standardization
IT	Information Technology
OEM	Original Equipment Manufacturer
POP	Purpose-objectives-premise
Q and A	Questions and Answers
QA	Quality Assurance
R&D	Research and development
RAM	Resource allocation matrix
ROI	Return on investment
SI	System integration
SIT	Stay in touch
T&Cs	Terms and conditions
TANSTAAFL	There ain't no such thing as a free lunch!
TEAM	Tenuous employee activities indicating mismanagement
TLC	Tender loving care
TQM	Total Quality Management
VADs	Value-Added Distributors
VARs	Value-Added Resellers

Index

C

D

G

H

I

J

K

S